Engineering and Costs of Dual Water Supply Systems

Engineering and Costs of Dual Water Supply Systems

S.L. Tang, Derek P.T. Yue and
Damien C.C. Ku

Publishing
LONDON • SEATTLE

Published by IWA Publishing, Alliance House, 12 Caxton Street, London SW1H 0QS, UK

Telephone: +44 (0) 20 7654 5500; Fax:+44 (0) 20 7654 5555; Email: publications@iwap.co.uk
Web: **www.iwapublishing.com**

First Published 2007

Copyedited and typeset by TechBooks, India

Cover design by Designforpublishing.co.uk

Index by Indexing Specialists (UK) Ltd

British Library Cataloguing in Publication Data
A CIP catalogue record for this book is available from the British Library

Library of Congress Cataloguing-in-Publication Data
A catalog record for this book is available from the Library of Congress
Library of Congress Control Number: 2006937261
ISBNs: 1843391325 and 9781843391326

Contents

1

Introduction to dual water supply systems

Hong Kong is the only place in the world that uses dual water supply systems extensively and on a large scale. The content of this book is therefore mainly based on the Hong Kong experience.

1.1 WHAT ARE DUAL WATER SUPPLY SYSTEMS?

The dual water supply systems involve two distribution systems: a fresh water system for potable use and another system conveying water for non-potable use, for example toilet flushing. The non-potable water can be seawater, raw (untreated) fresh water or reclaimed water (treated effluent) from sewage treatment works.

Very few coastal cities in the world use dual water supply systems, although these systems were operated satisfactorily in several locations worldwide (DFID 1997). A study on the effectiveness of using seawater for non-potable purpose and how the consumption of potable water can be reduced was carried out in 1996 (Smith *et al.* 1996). This study concluded that dual water supply systems could become an increasingly viable option for large urban areas in coastal zones. Hong Kong is one of those very few cities using the dual systems and is the only city that uses such systems in an extensive scale (Tang 2000). The dual systems have been available since 1950s in Hong Kong. The objectives of this book are to describe the

 ISBN: 9781843391326. Published by IWA Publishing, London, UK.

implementation of the dual water supply systems and evaluate the effectiveness of the systems technically and economically.

1.2 FRESH WATER SUPPLY SYSTEM

1.2.1 The Distribution System

Hong Kong is a coastal city in China with an area of about 1000 km^{2}and a population of 7 million. The mountainous nature of the territory has made the collection and storage of rainwater extremely difficult. The average annual fresh water demand in Hong Kong is about one billion m^3. As a result of high demand and low supply, about 70–80% of the fresh water used in Hong Kong is purchased from Guangdong Province of China (WSD 2005). The raw fresh water is extracted from Dongjiang (i.e. East River) at a point about 83 km away from the northern boundary of Hong Kong and is received at the Muk Wu Pumping Station located at the very north of the Hong Kong territory. From Muk Wu, the raw fresh water is delivered to the 17 storage reservoirs at various locations (i.e. 17 impounding reservoirs in which a total of about 30% of water demanded annually in Hong Kong is stored). Then, a series of treatment processes is carried out and after that, water is delivered to the consumers through a distribution system (see Figure 1.1).

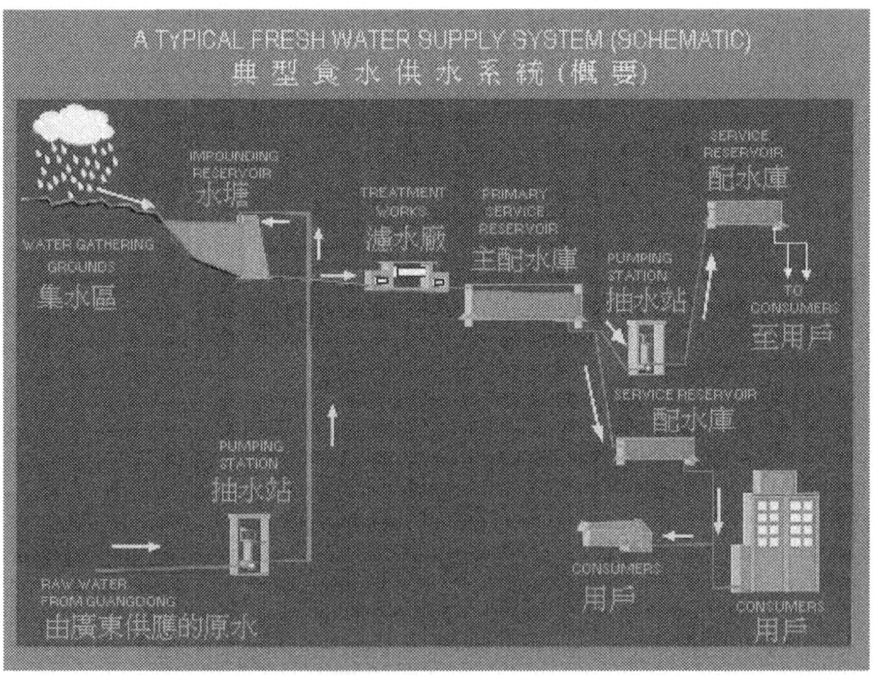

Figure 1.1 Fresh-water distribution system. *Source*: www.wsd.gov.hk. Accessed on June 2006.

The distribution system consists of about 170 service reservoirs situated throughout Hong Kong with a total capacity of about 3.5 Mm^3 (Mm^3 = million cubic metre), about 150 pumping stations with a total pumping capacity of about 30 Mm^3/day and a huge piping network consisting of a total of about 5500 km water mains. Water from the service reservoirs is distributed to consumers by gravity via extensive networks of water mains. The pressure in the system is sufficient to provide a direct supply to buildings of seven storeys above street level. With the assistance of individual pumping systems installed inside tall buildings, consumers at higher floor can be served. For remote village areas, the pressure is normally sufficient to provide a direct supply to three storeys above ground level.

1.2.2 Water Treatment Process

The usual water treatment processes adopted are firstly mixing and coagulation (with alum) in mixing chambers, then flocculation and sedimentation in clarifiers and then rapid gravity filtration. The filtered water passes into contact tanks where dosed chlorine and lime are added to disinfect and control the alkalinity of the final treated water. Fluoride is also added to minimise the risk of dental decay. A small amount of residual chlorine is maintained in the water to keep it free from bacteria on the rest of its journey. The final water is pumped into a system of pipelines, stored in service reservoirs and then supplied to the public. Figure 1.2 shows the treatment process diagrammatically (Hong Kong Water Supplies Department). This diagram shows the typical conventional treatment process. Advanced water treatment technologies, such as dissolved air floatation, ozonation and biological filtration, have been adopted in some new treatment plants. The quality standard for the potable water supply can fully comply with the WHO's guideline for drinking water quality.

There are 21 treatment plants at various locations in Hong Kong with a total capacity of treating 4.8 Mm^3 of fresh water per day. The treated fresh water will then enter into the distribution system. Water samples are collected at various locations of the entire system for water quality testing. The locations include catchment intakes, pumping stations, treatment works, service reservoirs, other strategic points in the distribution system and consumers' taps.

1.3 SEAWATER SUPPLY SYSTEM

About 80% of the population in Hong Kong is supplied with seawater for toilet flushing. People who are served with seawater usually live near the sea, mainly surrounding the Victoria Harbour (see shaded area in Figure 1.3).

The seawater supply system consists of pumping, treatment, storage and distribution. Seawater pumping stations are constructed near the shore, and the seawater is extracted from the sea directly. The treatment processes are much simpler, comprising only screening and disinfection, and they are usually carried out inside the pumping stations. The treated seawater is then pumped to service reservoirs

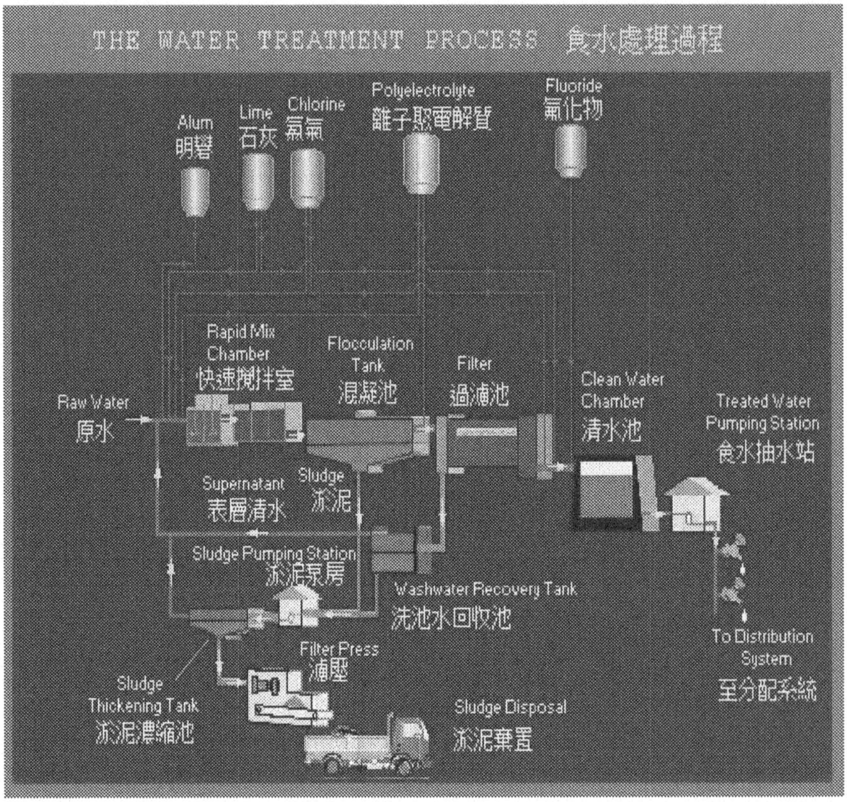

Figure 1.2 The fresh-water treatment process. *Source*: www.wsd.gov.hk. Accessed on June 2006.

and then to consumers by means of extensive pipeline reticulation systems (see Figure 1.4).

There are about 40 seawater pumping stations situated at various locations (usually at seaside) in Hong Kong, and their total pumping capacity is 1.5 Mm^3/day. There are also about 45 service reservoirs built for seawater, and their total capacity is about 0.2 Mm^3, with a network of seawater mains whose total length sums up to about 1400 km (WSD 2002). In order to avoid corrosion, seawater mains are normally made of materials such as polyethylene or polyvinyl chloride pipes.

1.4 SUMMARY

The dual water supply systems have been introduced in this chapter. Such water supply systems have been successfully run for 50 years in Hong Kong. The use of seawater for toilet flushing has undoubtedly helped reduce a considerable amount of fresh-water demand in Hong Kong. Provided that fundamental precautions

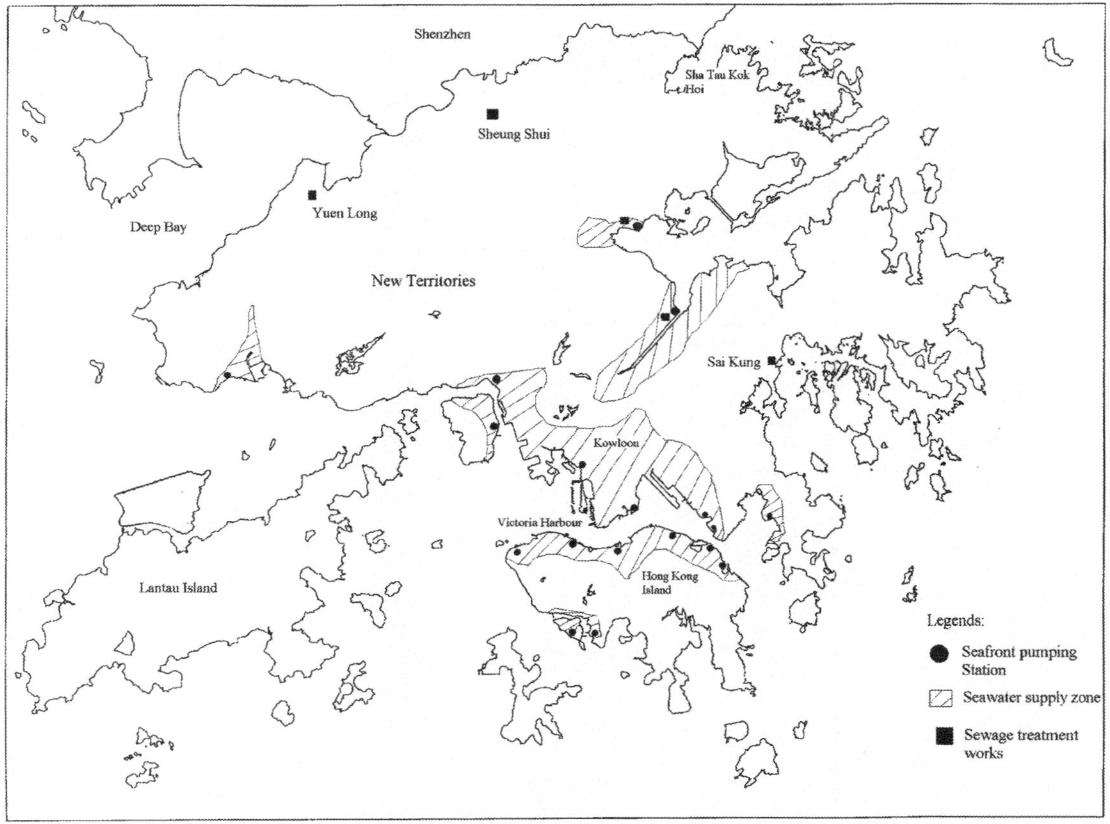

Figure 1.3 Areas (shaded) supplied with seawater for toilet flushing in Hong Kong.

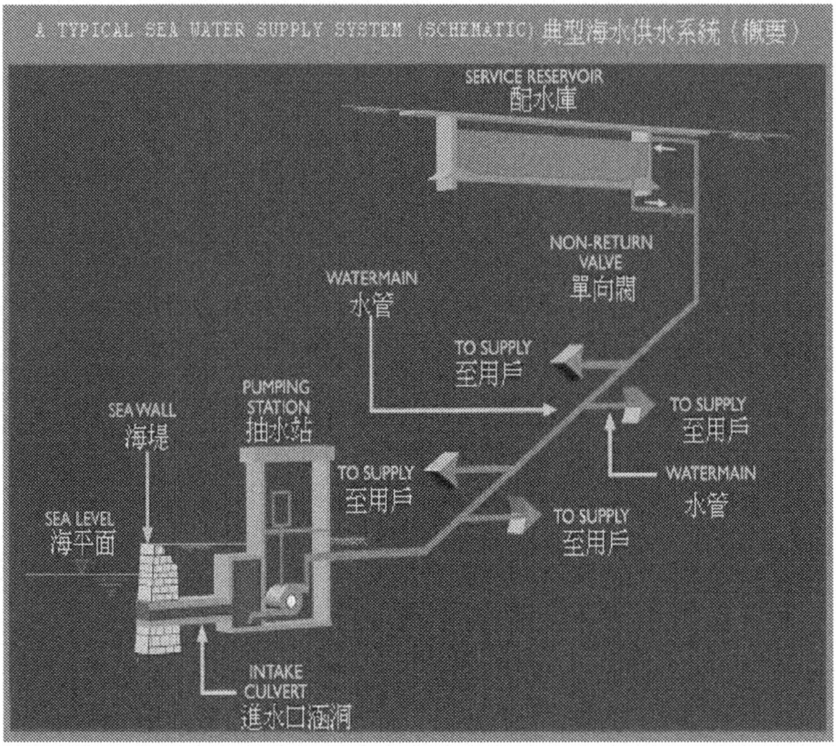

Figure 1.4 Seawater distribution system for toilet flushing. *Source*: http://wsd.gov.hk. Accessed on June 2006.

regarding the choice of materials are taken, a seawater supply system is technically not much difficult to construct, operate and maintain than any other reticulation systems.

There are problems and difficulties encountered in operating dual water supply systems, of course. Examples are the generation of saline wastewater, which is more difficult to treat in sewage treatment works, the easy corrosion of pipelines and equipment and so on. Chapter 2 will discuss in detail how these problems can be solved or mitigated. Besides seawater, the dual system may carry untreated raw fresh water or reclaimed water (treated effluent) from sewage treatment works instead of seawater to flush toilets. All these can help save potable water. They will be further discussed in later chapters.

1.5 REFERENCES

Department for International Development (DFID). 1997, Water – Saline sewerage and sewage treatment in coastal areas. Issue 4, May 1997, p. 5.

Smith, M. D., Stear, R. M. and Parr, J. 1996, Seawater for non-potable uses. Proceedings of the 22nd WEDC Conference, New Delhi, India, September, 1996, 241–243.
Tang, S. L. 2000, Dual water supply in Hong Kong. Proceedings of the 26th WEDC Conference on Water, Sanitation and Hygiene, Dhaka, Bangladesh, November, 2000, 364–366.
Water Supplies Department (WSD) 2002, Key Facts. Printing Department of Hong Kong Government, Hong Kong.
Water Supplies Department (WSD) 2005, Hong Kong's Water. Hong Kong Government Publication, Hong Kong.

2

Problems associated with operating a seawater distribution system

2.1 ADVANTAGES AND DISADVANTAGES OF DUAL WATER SUPPLY SYSTEMS WITH POTABLE WATER AND SEAWATER

The use of seawater for toilet flushing can offer many advantages to the water supply industry in Hong Kong. Seawater is an unlimited water resource with no cost. It helps save the use of fresh water, which is becoming a valuable and scarce commodity not only in Hong Kong but also in the whole world. The running cost of a seawater system is also low, and this will be discussed in Chapter 3. The quality of flushing water in Hong Kong is good in general, and very few complaints were reported in the past, except those related to aesthetic aspects such as odour and appearance. For most of the time, the quality of the seawater supplied to end users meets the requirements set by Water Supplies Department (WSD) of Hong Kong. The requirements are shown in Table 2.1 (Lee and Yu 1996). Compared with the criteria in Japan (Ogoshi *et al.* 2000) on using reclaimed wastewater for toilet flushing, the seawater criteria in Hong Kong are more stringent. It covers nine parameters in different aspects to ensure the aesthetic quality and health standard.

Table 2.1 Quality objectives/criteria for flushing

Parameters	Water quality objectives in Hong Kong		Criteria in Japan
	At intake point	At distribution system	
Colour (H.U.)	< 20	< 20	Note 1*
Turbidity (N.T.U.)	< 10	< 10	—
Suspended solids (mg/L)	< 10	< 10	—
Threshold odour number	< 100	< 100	Note 2†
Ammonia nitrogen (mg/L)	< 1	< 1	—
Dissolved oxygen (mg/L)	> 2	> 2	—
BOD_5 (mg/L)	< 10	< 10	—
Synthetic detergents (mg/L)	< 5	< 5	—
Escherichia coli (cfu/100 mL)	< 20000	< 1000	≤ 1000
pH	—	—	5.8–8.6

*Appearance – not unpleasant.
†Odour – not unpleasant.

However, there are also disadvantages in operating a seawater system. The problems include

- chlorination
- cross connection between the two supply systems
- deposits and biological growth inside pipelines
- corrosion
- sudden deteriorated quality of seawater
- ecology problems and
- treatment of mixed (salt + fresh) sewage.

These problems are to be addressed one by one in the following sections.

2.2 CHLORINATION PROBLEMS

Seawater for toilet flushing is abstracted at the intake point of seafront pumping stations. Inside the pumping stations, the seawater then passes through cross screens of mesh (size 5–10 mm) for debris removal and the chlorine generated from electro-chlorination is dosed to the screened water so as to control the biological growth. It can be argued that there may be potential problems that many seafront pumping stations are in the urban areas, and some people once had the worry that hydrogen, which is the by-product of the on-site chlorine generation, might impose explosion risk to the surroundings (Li *et al.* 2005).

In Hong Kong, an electro-chlorination technique has been widelyapplied for disinfection of such flushing water instead of conventional chlorination process, since seawater contains a high amount of sodium chloride. This electro-chlorination process is the on-site production of sodium hypochlorite (NaOCl) by passing an electric current through saltwater (electrolysis) using an electrolyser equipped with carbon steel cathodes and activated titanium anodes of monopolar design (Tang, S. L., Yue, Derek, P. T. and Li, X. Z.). The history of several decades' operation has convinced the public that electro-chlorination system is safe even in a congested area, such as Hong Kong.

2.3 CROSS CONNECTION PROBLEMS

The possibility of cross connection between the two supply systems is another disadvantage of using non-potable water for toilet flushing. WSD has a set of standard procedures to check against cross connection in providing seawater supply to consumers' premises. Nevertheless, subsequent modifications of the plumbing systems may cross connect the two supply systems unintentionally. Fortunately, as seawater has high salinity, consumers can easily detect any cross connection at an early stage, thus avoiding big problems to occur. A few incidents of cross connection were noted in the past. So far, the most unforgettable cross connection occurred in 1997. Due to the fault of construction site workers, the residents of two public housing estates in Ma On Shan were affected and some of them unfortunately used flushing water for cooking (Ho 2001). However, this kind of happenings has been very rare in the past.

2.4 DEPOSITS AND BIOLOGICAL GROWTH INSIDE PIPELINES

Deposits and biological growth in the seawater distribution system is another issue that needs to be carefully managed. It may lead to supply suspension and complaints on aesthetic quality. This can be minimised by effective maintenance on the electro-chlorination plant to ensure the uninterrupted chlorine dosing for marine growth control. Regular flushing of the distribution network to reduce accumulation of deposits and regular cleansing of uncovered service reservoir for better control of algal growth are important. In the past, the most serious incident that drew public attention occurred in 2001. The mussels grown in the flushing supply system blocked the pipes and caused supply suspension of some buildings. Again, this kind of happenings, however, has been very rare.

2.5 CORROSION PROBLEMS

High concentration of chloride in seawater can deepen the corrosion problems in the pipe network system. As shown in Table 2.2, the frequency of pipe burst for seawater flushing system is three times higher than that of fresh water system (WSD 2002). The frequency of supply suspension for seawater supply is therefore higher than that for the fresh water system. To tackle this, WSD has started to use

Table 2.2 Pipe burst of fresh and seawater supply networks

	Supply networks (km)		No. of main bursts repaired		No. of main burst/km-year	
Financial year	Fresh water	Seawater flushing	Fresh water	Seawater flushing	Fresh water	Seawater flushing
1997/98	4809	1111	760	713	0.16	0.64
1998/99	5127	1158	922	1029	0.18	0.89
1999/2000	5206	1227	884	938	0.17	0.76
2000/01	5458	1381	904	1029	0.17	0.75
2001/02	5574	1400	916	969	0.16	0.69

polyethylene pipe in 2000 for sizes less than 150-mm diameter (WSD 2001) to relieve the problem of pipe bursting.

Leaking seawater in a building will also accelerate deterioration of concrete and metal window frames. The supply tanks are often silted up with deposits carried by the seawater. Cleansing is a messy job but has to be done frequently.

2.6 SUDDEN DETERIORATION OF SEAWATER QUALITY

The quality of seawater supply depends very much on the marine water quality. The treatments provided at pumping stations are unlikely to give rise to great improvement or better control on the flushing water quality and, therefore, the quality of flushing supply is less stable than that of fresh water. Complaints may thus arise on any sudden quality deterioration. In addition, as the quality of treated seawater is inferior to that of potable water, stains are more easily accumulated in toilet bowls when seawater is used for toilet flushing.

Although the marine water quality in Hong Kong meets the quality requirements set for toilet flushing most of the time, it is worthy to note the rising trend of the turbidity, which may exceed the said requirements in the years to come and may deter the flushing water quality in the future. The existing treatment processes are not able to guarantee the turbidity of the flushing water to comply with the requirements. Hence, remedial actions should be taken for improving the deteriorating marine quality and precautionary measures (e.g. silt curtain at the intake) should be considered for use in the treatment processes in order to safeguard the quality of flushing water (Li *et al.* 2005).

2.7 ECOLOGY PROBLEMS

At present, in the rural inland areas of the New Territories of Hong Kong, sewage (without salinity) is treated by septic tanks and then discharged to rivers. These rural inland areas are far away from the coastal regions with a population of about 20% of total population in Hong Kong. With the possible introduction of seawater for toilet flushing in these areas, the sewage will have high conductivity of about 10,000 μS/cm. If this sewage is not collected by sewers, and because of its high salinity, the ecology of the rivers receiving it may be much disturbed. The survival of vegetation and aquatic life would be adversely affected. In addition, there may also be an increase of salinity on the farmland, and the affected farmland may no longer be suitable for agricultural purposes. Furthermore, even if all sewage is intercepted to sewage treatment works (STWs), the effluent (with salinity) is still eventually discharged to the rivers in these parts of Hong Kong. Therefore, the impacts due to the increase in salinity in the rivers should not be ignored as water in the rivers may no longer be suitable to be abstracted for potable or irrigation purposes. Hence, seawater supply for toilet flushing may only be suitable for areas where STWs are built such that the effluent coming out from these STWs is discharged back to the sea (Li *et al.* 2005).

2.8 TREATMENT OF MIXED (SALT + FRESH) SEWAGE

The used toilet flushing water (saline wastewater) is discharged into the sewerage system that leads the mixed (fresh + salt) wastewater going into the STW. The salt concentration of the mixed wastewater is between 5000 and 6000 mg/L – about one-fifth of seawater-salt concentration – in Hong Kong. Hence, the conventional method of treating wastewater has to be modified to suit the salinity added in the wastewater. The following sections describe the methods of treating mixed sewage (Tang and Lee 2002).

2.8.1 Sewage Treatment System Used in Hong Kong

The incoming sewage is introduced at the screw pumping station of the inlet works. It is then pumped to the screening chamber where coarse solids in the sewage are removed by bar screens. The screened sewage then goes to the detritors for the removal of grits. Then this sewage goes to the primary sedimentation tanks for settling suspended solids (see Figure 2.1).

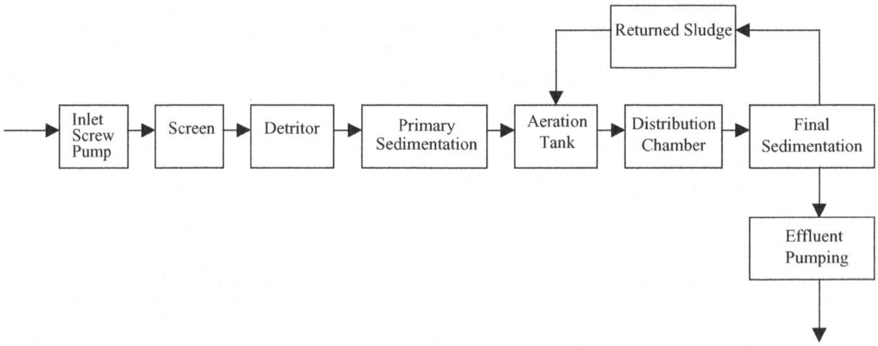

Figure 2.1 Typical STW schematic flow diagram in Hong Kong.

The settled sewage from the primary sedimentation tanks then goes to the aeration tanks for biological treatment. By the use of fine bubble diffusers, low-pressure air is fed continuously into the aeration tank to provide the oxygen essential for sustaining the life of the microorganisms, which assimilate the pollutants in the sewage. Extended aeration systems are used in the biological treatment processes so that the rear part of the long aeration tank is used for nitrification, and then later for de-nitrification (an anoxic process) with much less air supply at this part of the aeration tank.

The treated sewage then goes to the final sedimentation tanks in which the activated sludge is settled out. A controlled amount of this settled activated sludge is re-circulated to the aeration tank. The remaining activated sludge (surplus activated sludge) together with the primary sludge is then led to sludge treatment facilities in the STW.

The final effluent from final sedimentation tanks is then pumped through effluent pumping station to the sea outlet for disposal after disinfection.

2.8.2 Previous Laboratory Scale Research Work for Treating Mixed Sewage

About 30 years ago, an experiment (believed to be the first experiment of this sort) was carried out by Kessick and Manchen (1976) on the treatment of saline wastewater. In that experiment, with the addition of sea salts (39,100 mg/L) to primarily treated domestic wastewater, total suspended solids and volatile suspended solids were found to increase by 40% and 11%, respectively. The marked increase of suspended solids was believed to be because of the inhibition of the biodegradation of the suspended fraction in saline environment. Biodegradation of the soluble fraction, however, was found to be unaffected by sea salts added to levels expected in a seawater waste system.

Hamoda and Al-Attar (1995) carried out a study on the effects of high NaCl concentrations on activated sludge treatment. The findings were that the activated sludge process did not deteriorate as a result of the constant application of NaCl up to 30,000 mg/L, NaCl did not inhibit biomass growth, the salt appeared to stimulate the aggregation of microbial cells and improve sludge settling, and the response of the acclimated activated sludge saline-water system was similar to that of the fresh-water system and kinetic models developed for the fresh wastewater system could be used successfully. This phenomenon was most likely contributed by the growth of halophilic (i.e. salt-loving) microorganisms in the system.

For nitrification and de-nitrification, Dincer and Kargi (1999) carried out experiments on salt inhibition effects of these two processes. It was found that nitrification efficiency dropped markedly (to one-half or one-third) as salt levels increased and salt inhibition effects became more significant at salt concentrations above 20,000 mg/L. It was also found that nitrobacter (responsible for converting nitrite to nitrate) was more adversely affected than nitrosomonas (responsible for converting ammonia to nitrite), thus resulting in an accumulation of nitrite in the effluent at high salt concentrations. As to de-nitrification, it was found that when salt concentration was increased to a level above 10,000 mg/L, the de-nitrification efficiency dropped. De-nitrification rate seemed to be more sensitive to salt inhibition as compared to nitrification.

2.8.3 Experience of Treating Mixed Sewage in Tai Po STW in Hong Kong

In the aeration tank of activated sludge system in Tai Po STW, the heterotrophic organisms are responsible for BOD removal under the aerobic process. It is evident that a high BOD removal was found in the effluent, and it showed that there is no adverse effect on heterotrophic organisms under saline condition. There is also no significant effect on COD removal found in such a condition. It was supported by the paper written by Yu *et al.* (2000) that the COD removal efficiency was not affected in saline conditions in some bench-scale experiments.

Experience in Tai Po STW is that the nitrification process has been inhibited in saline wastewater. By the way, from the paper written by Yu *et al.* (2000) bench-scale investigations of nitrification have been done and it was found that nitrification is greatly impaired with increase in salinity. As the degree of salinity fluctuates throughout a day, the variation of Cl^- ions in wastewater due to different peak hours of people using toilet flushing considerably affects the rate of nitrification in the treatment process. The activity of autotrophic organisms responsible for nitrification was determined in several batch tests and was found to be about one-half to one-third of the degree of activities observed in fresh wastewater.

In order to restore the achievement of the original full nitrification for activated sludge nitrogen removal process, the Tai Po STW in Hong Kong has the sludge age and mixed liquor suspended solids (MLSS) increased in order to compensate for this deficiency. Sludge age was increased from 10 days in fresh wastewater to 15 days in saline wastewater. At the same time, the MLSS is increased from 2000 mg/L in fresh wastewater concentration to approximately 3000 mg/L in saline conditions. When given sufficient nitrifying bacteria in the system, there is no need to lengthen the aeration time and the amount of oxygen required. The treatment plant just keeps the same hydraulic retention time, which is about 8–11 hours, and with the same dissolved oxygen level maintained at around 2 mg/L before and after the introduction of saline wastewater.

The de-nitrification process, unlike BOD/COD removal and nitrification processes (which require a large amount of oxygen), is basically anoxic. The result found in Tai Po STW is that the heterotrophic organisms in saline wastewater perform satisfactorily in the de-nitrification process even the wastewater is saline. The presence of dissolved oxygen is measured by redox potential, which is between +50 mV and −100 mV. This means that the wastewater contains nitrite ions and the anoxic process does occur. It appears that no adverse effect is found on de-nitrification in saline wastewater in Tai Po STW.

2.8.4 Sulphate Concentration and Odour Problems

The introduction of seawater to wastewater treatment plants not only adversely affects nitrification but also increases the sulphate concentration of biological processes because seawater contains considerably undesirable sulphur compounds. The increase of sulphate concentration can lead to an increase of H_2S in any down-stream anaerobic process such as sludge digestion. In order to reduce the release of H_2S gases during biological treatment processing, a certain amount of ferric chloride (Fe_2Cl_3) is added. However, the added Fe_2Cl_3can create corrosion problems in tanks, pipelines and so on. So, ferric chloride chemical must be used with caution, and the tanks and pipelines must be built with strong materials. Besides odour problems, a marked reduction of ortho-phosphate (O-P) content in sewage can be found when sulphate concentration increases. The reduction of soluble phosphate is believed to be a chemical precipitation process associated with the low solubility of most phosphate salts. It is then removed as primary sludge in the primary sedimentation process. Therefore the level of O-P content must be monitored carefully in order to ascertain that the ratio of BOD:N:P is 100:5:1, as required by the Drainage Services Department of Hong Kong. If the required P is reduced in saline condition, phosphate acid will be added so that the O-P concentration can be increased to reach the required standard.

2.9 SUMMARY

The addition of seawater to the fresh wastewater, from the experience obtained in Hong Kong, does not significantly affect BOD/COD removal efficiency and de-nitrification. Salinity in wastewater affects mainly nitrification. An increase of the concentration of the MLSS from 2000 to 3000 mg/L and the sludge age from 10 to 15 days can improve nitrification in saline condition. This can be achieved by reducing the wastage of the returned activated sludge (i.e. the settled activated sludge from the final sedimentation tanks) and increasing its circulation back to the aeration tank.

2.10 REFERENCES

Dincer, A. R. and Kargi, F. 1999, Salt inhibition of nitrification and denitrification in saline waste water. Environmental Technology 20, 1147–1153.

Hamoda, M. F. and Al-Attar, M. S. 1995, Effects of high sodium chloride concentrations on activated sludge treatment. Water Science Technology 31[9], 61–72.

Ho, P. Y. 2001, Water for a Barren Rock – 150 Years of Water Supply in Hong Kong. The Commercial Press, Hong Kong.

Kessick, M. A. and Manchen, K. L. 1976, Salt water domestic waste treatment. J. WPCF 48[9], 2131–2136.

Lee, C. K. and Yu, C. W. 1996, Conservation of water resources – use of sea water for flushing in Hong Kong. 10th IWSA-ASPAC Regional Conference and Exhibition, Hong Kong, November 1996.

Li, X. Z., Luk, S. F. and Tang, S. L. 2005, Sustainability of toilet flushing water supply in Hong Kong. Journal of CIWEM, 19[2], 85–90.

Ogoshi, M., Suzuki, Y. and Asano, T. 2000, Non-potable Urban Water Reuse – a case of Japanese water recycling. Water 21, 27–30.

Tang, S. L. and Lee, T. H. 2002, Treatment of mixed (fresh + salt) wastewater. Proceedings of the 28th WEDC Conference, Calcutta, India, November 2002, 56–58.

Tang, S. L., Yue, D. P. T. and Li, X. Z. In press, Comparison of engineering costs of seawater, reclaimed water and fresh water for toilet flushing in Hong Kong. Journal of CIWEM 20[3].

Water Supplies Department (WSD). 2001, Key Facts. Printing Department of Hong Kong Government, Hong Kong.

Water Supplies Department (WSD). 2002, Annual Report 2001–2002. Printing Department of Hong Kong Government, Hong Kong.

Yu, S. M., Leung, W. Y., Ho, K. M., Greenfield, P. F. and Eckenfelder, W. W. 2000, The Impact of Seawater Flushing on Biological Nitrification-Denitrification Activated Sludge Sewage Treatment Process. Drainage Services Department, Hong Kong Government, Hong Kong.

3

Cost comparison of dual water supply systems with conventional single (potable) fresh water system

In order to compare the engineering economy of the existing dual water supply systems with a fresh water supply system, an imaginary single fresh water supply system with a similar size to the existing dual systems has to be assumed. The consumption of saltwater for flushing is to be replaced by fresh water. The extra costs in procuring the extra raw fresh water and in treating this amount of raw fresh water must be taken into account. The saving in this imaginary case is the costs for construction and maintenance of the extra distribution system for seawater (Li *et al.* 2005).

The capital investments can be divided into initial capital costs and the renewal costs of the water work facilities. Besides capital investments, there are running costs, which are recurrent expenditures spent every year by the Water Supplies Department (WSD) of Hong Kong in order to have the water supply system successfully operated. On the other hand, the WSD receives incomes every year from the water users. The capital investments, running costs and incomes are discussed in more details next.

3.1 CAPITAL INVESTMENTS

For the fresh water supply system, the initial capital includes the construction cost of water work facilities, the cost of purchasing equipment and the land cost. However, for the case of Hong Kong the lands on which the facilities are built are owned by the Hong Kong Government, and since the WSD is also owned by the Hong Kong Government, we therefore ignore the land cost in this analysis. The water work facilities include impounding reservoirs, water treatment plants, pumping stations, service reservoirs, E&M equipment and water mains. For the seawater supply system, the facilities include only pumping stations, service reservoirs, E&M equipment and seawater mains. There are no impounding reservoirs and treatment plants in the seawater supply system.

Different facilities have different renewal costs, which in turn relate to depreciation. The depreciation of the facilities is calculated on a straight-line basis to write off the costs of the facilities over their estimated useful lives. The annual rates of depreciation are set as 1% for tunnels and dams, 2% for civil engineering works and fresh water mains, 5% for saltwater mains, 4% for E&M items and 10–20% for motor vehicles (WSD 2002b). The depreciation rates relate to the lifetimes of different facilities, which indicate when the renewal costs of individual facilities will have to be incurred. However, from the engineers' experiences of the WSD, there is no exact lifetime for water mains. This is because the mains are usually replaced before the end of their lifetime due to system expansion or work improvement, or because they are destroyed by the construction of underground utilities or failure of road bases. The same happens to other facilities (e.g. E&M items) as well. Nevertheless, on a conservative side, we assume the lifetimes of the saltwater mains and the fresh water mains to be 20 years and 30 years, respectively. Both distribution systems, as told by the engineers of the WSD, use the same type of pipe materials and the mains of the two systems are of the same cost per unit length.

3.2 RUNNING COSTS

Running costs include (1) staff cost, (2) operation and administration expenses (OAE) and (3) the cost of purchasing raw water from China. About 70% of the total fresh water usage in Hong Kong is supplied by China and a cost is paid for the raw (untreated) water purchased. In 2001–2002, the staff establishment of the WSD was over 5000, which costs HK$1747 million (US$1 = HK$7.8) in that year (WSD 2002a). OAE are the costs of light and power; the maintenance costs of buildings, reservoirs, water mains, treatment plants and pumping stations; the costs of emergency repairs, hiring of services, professional fees, specialist supplies, material and equipment; and general departmental expenses. The OAE totally amounted to HK$1485 million in 2001–2002 (WSD 2002a). The water purchasing cost for 2001–2002 was HK$2445 million. The total running cost in that year was therefore HK$5677 million (i.e. 1747 + 1485 + 2445). In the

Table 3.1 Staff cost of the existing dual water supply systems

Year	CPI	Staff cost (HK$ million)	Real staff cost at 2000 (HK$ million)	Fresh water consumption (M m^3)	Product
1992/93	72.0	878	1,219	889	1,084,086
1993/94	78.3	1,013	1,293	915	1,183,773
1994/95	85.2	1,118	1,312	923	1,211,166
1995/96	92.9	1,275	1,372	919	1,261,275
1996/97	98.8	1,396	1,412	928	1,311,222
1997/98	104.5	1,474	1,410	913	1,287,810
1998/99	107.5	1,603	1,491	916	1,365,905
1999/2000	103.2	1,679	1,626	911	1,482,140
2000/01	99.4	1,734	1,744	924	1,611,887
2001/02	97.8	1,747	1,786	940	1,679,120
		Total	14,670	9,178	13,478,389
				Weight mean	1,468

Source: WSD annual report (2002a).
CPI = consumers' price indices.
Real cost = cost at year 2000 money value after adjustment by CPI (see Section 3.5).

economy comparison in the later part of this section, we do not use the figure for a particular year (e.g. HK$5677 million) but an average figure from the past 10 years. Table 3.1 shows the average figure of staff cost calculated from 10 years' records, which is equal to HK$1468 expressed at year 2000 money value. Table 3.2 shows

Table 3.2 The OAE of the existing dual water supply systems

Year	CPI	OAE (HK$ million)	Real OAE (HK$ million)	Fresh water consumption (M m^3)	Product
1992/93	72.0	747	1,037	889	922,337
1993/94	78.3	832	1,063	915	972,261
1994/95	85.2	978	1,148	923	1,059,500
1995/96	92.9	1,177	1,267	919	1,164,330
1996/97	98.8	1,147	1,161	928	1,077,344
1997/98	104.5	1,247	1,193	913	1,089,484
1998/99	107.5	1,303	1,212	916	1,110,277
1999/2000	103.2	1,332	1,291	911	1,175,825
2000/01	99.4	1,330	1,338	924	1,236,338
2001/02	97.8	1,485	1,518	940	1,427,300
		Total	12,228	9,178	11,234,998
				Weight mean	1,224

Source: WSD annual report (2002a).
Real cost = cost at year 2000 money value after adjustment by CPI (see Section 3.5).

Table 3.3 The cost of raw water purchased for the existing dual water supply systems

Year	CPI	Raw water cost (HK$ million)	Real raw water cost (HK$ million)	Fresh water consumption (M m^3)	Product
1992/93	72.0	1,079	1,499	889	1,332,265
1993/94	78.3	1,200	1,533	915	1,402,299
1994/95	85.2	1,302	1,528	923	1,410,500
1995/96	92.9	1,541	1,659	919	1,524,412
1996/97	98.8	1,812	1,834	928	1,701,960
1997/98	104.5	2,007	1,921	913	1,753,484
1998/99	107.5	2,223	2,068	916	1,894,203
1999/2000	103.2	2,373	2,299	911	2,094,770
2000/01	99.4	2,425	2,440	924	2,254,225
2001/02	97.8	2,445	2,500	940	2,350,000
				9,178	17,718,119
				Weight mean	1,931

Weight mean (no wastage) $= 1,931 \times (1 - 5.77\%) = 1,819$

Source: WSD annual report (2002a).
Real cost = cost at year 2000 money value after adjustment by CPI (see Section 3.5).

the same for OAE, which is equal to HK$1224, and Table 3.3 shows that for raw water cost, which is HK$1819. The total running cost used for cost comparison in Section 3.5.1 is therefore HK$4511 million (i.e. 1468 + 1224 + 1819) at year 2000 money value.

3.3 INCOME GENERATED

The income received by the WSD, the only water supplier in Hong Kong, comes from water charges from users. The charging method is quite complicated and not to be discussed in detail here. In 2001–2002, the total water charges amounted to HK$5629 million (WSD 2002a). Similarly, in the economy comparison, we do not use this figure but the figure averaged from the past 10 years.

3.4 NET PRESENT VALUE

The net present value (NPV) will be used as the tool for cost comparison in the next section. Before going to the details of the cost comparison, NPV is to be introduced in this section.

Figure 3.1 shows that F would be equal to $16,289 if an interest rate i is taken as 5% per year. In other words, $16,289 of 10 years later from now is equivalent to $10,000 today. Or it can be said that the *present value* (i.e. today's value) of $16,289 of 10 years later is now $10,000. Mathematically, the present value of an

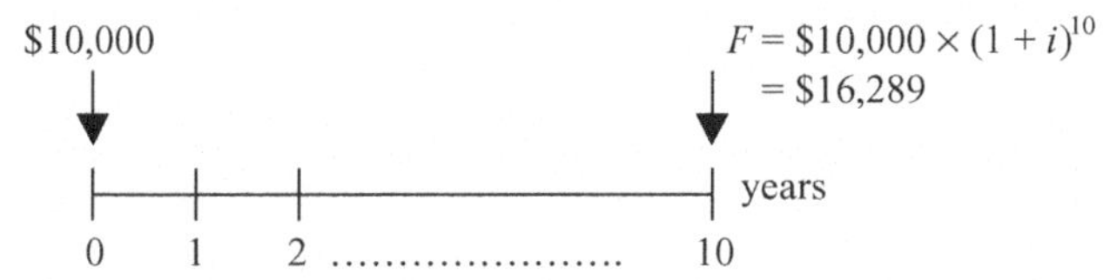

Figure 3.1 Present value of a single payment F.

amount F in general can be expresses in the form of

$$\text{Present value of } F = \frac{F}{(1+i)^n} = F \times \frac{1}{(1+i)^n},$$

where $n =$ the number of years from today in future.

3.4.1 An Example of Calculating NPV

Find the NPV of an investment of \$10,000 whose receipts in the next 3 years are \$4000, \$4500 and \$5000, respectively.

Solution

The calculation of NPV is best performed in a tabulated manner as shown in Table 3.4.

Column (1) and (2) are the cash flows given as data for finding the NPV. Column (3), the net cash flow column, is calculated by subtracting Column (1) from Column (2). Column (4) is a column of figures which are calculated using the formula $1/(1+i)^n$. Column (5), the discount cash flow, is a column of figures which are the multiples of Column (3) and Column (4). The NPV is the sum of all figures in Column (5).

The NPV is calculated to be \$2211, which is a positive value. Because the NPV is positive, this investment is profitable.

Table 3.4 Calculation of NPV

End of year	(1) Cash out	(2) Cash in	(3) = (2) − (1) Net cash flow	(4) $1/(1+i)^n$	(5) = (3) × (4) Discount cash flow
0	10,000		−10,000	1	−10,000
1		4,000	4,000	0.9524	3,810
2		4,500	4,500	0.9070	4,082
3		5,000	5,000	0.8638	4,319
					NPV = 2,211

Note: i is taken as 5% per year in this example.

Table 3.5 Calculation of NPV

End of year	(1) Cash out	(2) Cash in	(3) = (2) − (1) Net cash flow	(4) $1/(1+i)^n$	(5) = (3) × (4) Discount cash flow
0	30,000		−30,000	1	−30,000
1	1,000	3,500	2,500	0.9524	2,381
2	1,000	3,500	2,500	0.9070	2,268
3	1,000	3,500	2,500	0.8638	2,160
4	1,000	3,500	2,500	0.8227	2,057
5	1,000	3,500	2,500	0.7835	1,959
6	1,000	3,500	2,500	0.7462	1,866
7	1,000	3,500	2,500	0.7107	1,777
8	1,000	3,500	2,500	0.6768	1,692
9	1,000	3,500	2,500	0.6446	1,612
10	1,000	3,500	2,500	0.6139	1,535
					NPV = −10,693

Note: i is taken as 5% per year; all values are in thousand dollars.

3.4.2 A Further Example of Calculating NPV

The construction cost of a sewerage system in a small village is estimated to be \$30,000,000. The annual operation, maintenance and repair cost will be \$1,000,000 per year. The annual income from the collection of sanitation fees from the users will be \$3,500,000. Taking a time horizon of 10 years and a discount rate of 5%, determine if the project is viable.

Solution

The calculation of NPV is again performed in a tabulated manner as shown in Table 3.5.

It can be seen that the NPV is −\$10,693, a negative value. This means that the project is not profitable.

3.5 COST COMPARISON OF THE TWO SYSTEMS

The method of NPV will be used to compare the engineering economy of the existing dual water supply and the imaginary single fresh water supply systems. The time horizon is taken as 60 years. The NPV of each system can be calculated using the following formula (Tang 2003; Chapters 2 and 3):

$$\mathrm{NPV} = F_0 + \frac{F_1}{(1+i)^1} + \frac{F_2}{(1+i)^2} + \cdots + \frac{F_{60}}{(1+i)^{60}},$$

where F_n = 'net cash flow' or 'income − expenditure' in year n.

Expenditure includes, besides running cost for each year ($n = 1, 2, \ldots, 60$), the capital cost when $n = 0$ and the renewal cost when $n = 30$ for the fresh water system, or when $n = 20$ and 40 for the seawater system. E&M items are renewed when $n = 30$. Motor vehicles are renewed when $n = 5, 10, 15, \ldots, 55$. It is further explained next.

Since the seawater mains system is assumed to have a lifetime of 20 years and the fresh water mains system 30 years, we have to make a comparison based on a time horizon of 60 years. The renewal of the seawater mains will be at years 20 and 40 and that of the fresh water mains will be at year 30. All prices have to be adjusted to the money value of Year 2000 by the use of CPI published by the Hong Kong Government (CSD 2003). By doing so, the economy comparison can be made on a 'constant base year prices' basis, and hence the NPV calculated will be 'real' (Tang 2003; Chapter 4) without the necessity of inflation adjustment. The two alternatives will be compared using their NPVs. The alternative with higher NPV, or 'less negative NPV' in this case, is therefore a better alternative. The following section shows the data and the calculations.

3.5.1 Existing Dual Water Supply Systems

The capital cost, including fresh water impounding reservoirs, fresh water mains and pumping stations, fresh water treatment plants, fresh water service reservoirs, E&M equipment, saltwater mains and pumping stations, and saltwater service reservoirs, is HK$35,830 million (WSD 2002a). This cost has been expressed as a figure in year 2000 money value. Different facilities may be provided at different times in the past. The cost of each facility provided is adjusted to the Year 2000 price by applying CPI adjustment. Summing up the adjusted prices for all the facilities will arrive at the figure HK$35,830 million. Besides the capital cost, there are renewal costs for the fresh water mains and seawater mains. These renewal costs are calculated as shown in Table 3.6. Other costs can also be found in WSD (2002a). They are shown next after adjusted by CPI to year 2000 prices.

Renewal cost of fresh water mains (30 years life) = HK$21,608 million.

Renewal cost of saltwater mains and (20 years life) = HK$3942 million.

Renewal cost of E&M equipment and meters (30 years life) = HK$544 million.

Renewal cost of motor vehicles (5 years life) = HK$50 million.

Running cost (staff cost, OAE and water purchase cost) = HK$4511 million per year.

Income (water charges) = HK$5227 million per year.

The aforementioned figures are used in the NPV calculation, which is shown in Table 3.7. The NPV calculated, using a discount rate, i, of 5% per annum for

Table 3.6 Calculation of renewal cost of mains (pipe cost)

	Year basis	CPI	Length (km)	Pipe cost (HK$/m)	Real cost (HK$/m)	Total cost (HK$ million)
Fresh water mains	2000/01	99.4	5458	3,935	3,959	21,608
Seawater mains	2000/01	99.4	1381	2,837	2,855	3,942

Source: Mr. Wong, Senior Engineer of WSD (Head Quarter).
Real cost = cost at year 2000 money value after adjustment by CPI.

Table 3.7 NPV calculation for the existing dual water supply systems

Year	(1) Capital cost	(2) Running cost	(3) Income	(4) = (3) − (2) − (1) Net cash flow	(5) Discount cash flow at $i = 5\%$
0	35,830			−35,830	−35,830
1		4,511	5,227	716	682
2		4,511	5,227	716	649
3		4,511	5,227	716	618
4		4,511	5,227	716	589
5	50	4,511	5,227	666	521
6		4,511	5,227	716	534
7		4,511	5,227	716	509
8		4,511	5,227	716	484
9		4,511	5,227	716	461
10	50	4,511	5,227	666	408
11		4,511	5,227	716	418
12		4,511	5,227	716	399
13		4,511	5,227	716	380
14		4,511	5,227	716	361
15	50	4,511	5,227	666	320
16		4,511	5,227	716	328
17		4,511	5,227	716	312
18		4,511	5,227	716	297
19		4,511	5,227	716	283
20	3,942 + 50	4,511	5,227	−3,276	−2,816
21		4,511	5,227	716	257
22		4,511	5,227	716	245
23		4,511	5,227	716	233
24		4,511	5,227	716	222
25	50	4,511	5,227	666	196
26		4,511	5,227	716	201
27		4,511	5,227	716	192
28		4,511	5,227	716	183
29		4,511	5,227	716	174
30	21,608 + 544 + 50	4,511	5,227	−21,486	−4,972
31		4,511	5,227	716	158
32		4,511	5,227	716	150
33		4,511	5,227	716	143
34		4,511	5,227	716	136
35	50	4,511	5,227	666	121
36		4,511	5,227	716	124
37		4,511	5,227	716	118
38		4,511	5,227	716	112
39		4,511	5,227	716	107
40	3,942 + 50	4,511	5,227	−3,276	−466
41		4,511	5,227	716	97
42		4,511	5,227	716	92
43		4,511	5,227	716	88
44		4,511	5,227	716	84
45	50	4,511	5,227	666	74
46		4,511	5,227	716	76
47		4,511	5,227	716	72

Table 3.7 (*Continued*)

Year	(1) Capital cost	(2) Running cost	(3) Income	(4) = (3) − (2) − (1) Net cash flow	(5) Discount cash flow at i = 5%
48		4,511	5,227	716	69
49		4,511	5,227	716	66
50	50	4,511	5,227	666	58
51		4,511	5,227	716	59
52		4,511	5,227	716	57
53		4,511	5,227	716	54
54		4,511	5,227	716	51
55	50	4,511	5,227	666	45
56		4,511	5,227	716	47
57		4,511	5,227	716	44
58		4,511	5,227	716	42
59		4,511	5,227	716	40
60		4,511	5,227	716	38
					NPV = −29,627

the existing dual water supply systems, is found to be −HK$29,627 million. The negative NPV shows that running water supply to the people of Hong Kong is not a profitable business.

3.5.2 Imaginary Single Fresh Water Supply System

The capital cost (impounding reservoirs, water treatment plants, service reservoirs, E&M equipment, pumping stations, water mains (one distribution system instead of two but of larger pipe sizes)) is HK$37,691 million at year 2000 money value. This capital cost of the imaginary single potable fresh water system is obtained by firstly deducting the capital cost of saltwater components from all existing dual system facilities and then multiplying that figure by (917.8 + 181.6)/917.8, where 917.8 Mm^3 is the average annual fresh water consumption and 181.6 Mm^3 is the average annual seawater consumption in Hong Kong for the past 10 years.

Renewal cost of water mains (30 years life) = HK$21,608 × (917.8 + 181.6)/917.8 = HK$25,884.

Renewal cost of E&M equipment and meters (30 years life) = HK$544 million.

Renewal cost of motor vehicles (5 years life) = HK$50 million.

Running cost = HK$5133 million per year (see Tang, S. L., Yue, Derek, P. T. and Li, X. Z.) 2003, p. 60 and 64).

Income = HK$5283 million per year (see Tang, S. L., Yue, Derek, P. T. and Li, X. Z.) 2003, p. 60, 61 and 65).

A method similar to what has been shown in Table 3.7 is used to calculate the NPV for the imaginary single potable fresh water supply system, and is shown in Table 3.8. The NPV for the imaginary single system is calculated to be −HK$41,133 million in Table 3.8. Comparing with the NPV of −HK$29,627

Table 3.8 NPV calculation for the imaginary single potable fresh water supply system

Year	(1) Capital cost	(2) Running cost	(3) Income	(4) = (3) − (2) − (1) Net cash flow	Discount cash flow at $i = 5\%$
0	37,691			−37,691	−37,691
1		5,133	5,283	150	143
2		5,133	5,283	150	136
3		5,133	5,283	150	130
4		5,133	5,283	150	124
5	50	5,133	5,283	100	78
6		5,133	5,283	150	112
7		5,133	5,283	150	107
8		5,133	5,283	150	102
9		5,133	5,283	150	97
10	50	5,133	5,283	100	61
11		5,133	5,283	150	88
12		5,133	5,283	150	84
13		5,133	5,283	150	80
14		5,133	5,283	150	76
15	50	5,133	5,283	100	48
16		5,133	5,283	150	69
17		5,133	5,283	150	66
18		5,133	5,283	150	62
19		5,133	5,283	150	59
20	50	5,133	5,283	100	38
21		5,133	5,283	150	54
22		5,133	5,283	150	51
23		5,133	5,283	150	32
24		5,133	5,283	150	47
25	50	5,133	5,283	100	29
26		5,133	5,283	150	42
27		5,133	5,283	150	40
28		5,133	5,283	150	38
29		5,133	5,283	150	36
30	25,884 + 544 + 50	5,133	5,283	−426,329	−6,092
31		5,133	5,283	150	33
32		5,133	5,283	150	32
33		5,133	5,283	150	30
34		5,133	5,283	150	29
35	50	5,133	5,283	100	18
36		5,133	5,283	150	26
37		5,133	5,283	150	25
38		5,133	5,283	150	24
39		5,133	5,283	150	22
40	50	5,133	5,283	100	14
41		5,133	5,283	150	20
42		5,133	5,283	150	19
43		5,133	5,283	150	18
44		5,133	5,283	150	17
45	50	5,133	5,283	100	11
46		5,133	5,283	150	16
47		5,133	5,283	150	15

Table 3.8 (*Continued*)

Year	(1) Capital cost	(2) Running cost	(3) Income	(4) = (3) − (2) − (1) Net cash flow	Discount cash flow at $i = 5\%$
48		5,133	5,283	150	14
49		5,133	5,283	150	14
50	50	5,133	5,283	100	9
51		5,133	5,283	150	13
52		5,133	5,283	150	12
53		5,133	5,283	150	11
54		5,133	5,283	150	11
55	50	5,133	5,283	100	7
56		5,133	5,283	150	10
57		5,133	5,283	150	9
58		5,133	5,283	150	9
59		5,133	5,283	150	8
60		5,133	5,283	150	8
					−41,133

for the existing dual water supply systems, the single system is nearly 40% (or more exactly 39%) more expensive than the existing dual systems.

3.6 REFERENCES

Census and Statistical Department (CSD). 2003, Consumer Price Indices. The Gorvernment of Hong Kong SAR, Hong Kong.

Li, X. Z., Luk, S. F. and Tang, S. L. 2005, Sustainability of toilet flushing water supply in Hong Kong Water and Environment Journal, Journal of CIWEM, 19[2], 85–90.

Tang, S. L. 2003, Economic Feasibility of Projects: Managerial and Engineering Practice. 3rd edition, Chinese University Press, Hong Kong.

Tang, S. L., Tsang, W. L. and Leung, H. Y. 2003, A Comparative Study of the Economy of 'Single Water Supply System' and 'Dual Water Supply Systems'. Research Report, Civil and Structural Engineering Department, The Hong Kong Polytechnic University. (Available in the library of the Hong Kong Polytechnic University).

Water Supplies Department (WSD). 2002a, Annual Report 2001–2002. Printing Department of Hong Kong Government, Hong Kong.

Water Supplies Department (WSD). 2002b, Operating Accounts for the Year Ended 31 March 2002. Printing Department of Hong Kong Government, Hong Kong.

4

Using reclaimed water in the dual water supply systems

4.1 THE NEED

The use of reclaimed water, in one form or another, has long been a common practice all over the world in times of drought. When and where water resources are scarce, people will have no choice but to use as little water as possible, and will naturally save the less dirty water for toilet flushing, floor cleansing, etc. Such a practice is the most primitive way of segregating wastewater for reuse, and the percentage of recovery can be very high depending on the incentive of the consumers to save water, as impelled by the severity of water shortage. The only drawback is inconvenience.

Take Hong Kong as an example. Based on the 2002–2004 statistics of Water Supplies Department, 80% of the average annual water consumption of the territory was met by imported water. The local water resources are only able to meet around 20% of the demand. Hong Kong had long been suffering from water shortage until adequate raw water supply from the Dongjiang (East River) was secured by an agreement made between the governments of Hong Kong and the adjoining Guangdong Province, starting from the mid-1960s. Before that, water rationing was a norm; during the most severe periods, public water supply was available for only 4 hours every fourth day. It might not even have enough pressure to reach the

S. L. Tang, Derek Yue and Damien Ku. ISBN: 9781843391326. Published by IWA Publishing, London, UK.

upper floors of a building. Every household at that time kept stock of the precious fresh water in basins and buckets after initial use for later reuse in all practical ways.

However, once the threat of water rationing was lifted, very few people continued to restrict their ways of using water, not to say storing wastewater for reuse. Reclaimed water is even less attractive now in Hong Kong because seawater for toilet flushing is supplied free of charge, with heavy subsidy from the Government. In 2002, the provision of flushing water to new development areas and existing areas in the North-West New Territories was examined. It was found that the unit life-cycle cost of supplying seawater for flushing to a district of 740,000 population was HK\$3.5/m^3 (Ku *et al.* 2003). For consumers in districts where seawater supply is not yet available, a temporary fresh water supply connection is normally approved if applied for. Such a consumer account can enjoy a free supply of 30 m^3 fresh water every 4 months for toilet flushing, and over this amount the extra quantity used will be charged at HK\$4.58/m^3 which is acceptable to most consumers (WSD 2005).

Hence, the promotion of reclaimed water schemes must start with the incentive of the general public to use it. Such an incentive can be aroused by publicity programmes, but in the long run it can only be sustainable under the drive of legislations, obligations, economic and convenience considerations. In this regard, a motivated water and sewage charging mechanism supported by public opinion surveys and government water saving policies will be necessary.

Many reclaimed water supply projects have been implemented all over the world in the past few decades, mainly to optimise the use of scarce water resources. Basically, these projects are designed to recover used water, which will otherwise run to waste, treat the water and distribute it to individual households to meet the requirements of the consumers, in quantity and quality, for the purpose of the intended use. To receive public acceptance, these systems have also to be environment friendly, reliable and simple to operate at low capital and running costs, bearing in mind the need for dual supply systems.

Actually, apart from better utilisation of water resources, reclaimed water has the following added values, some of which cannot be achieved with the use of seawater for toilet flushing:

(a) reduction of wastewater volume for sewage treatment,
(b) maintaining a steady supply of water for landscape irrigation, street cleansing, air-conditioning, etc., thus keeping a pleasant living environment even in times of drought,
(c) replenishment of ground water or supplement to local catchment,
(d) arousing public awareness of the need for water conservation and
(e) promoting general water-saving habits.

4.2 QUANTITY CONSIDERATIONS

4.2.1 Quantity Requirement

The quantity requirement of a reclaimed water project depends on the intended use of the water. In view of potential health hazards and the difficulty in gaining public acceptance, reclaimed water is normally supplied for non-potable use only, such as toilet flushing, landscape features, irrigation, floor or street cleansing, car washing, water-cooled system for air-conditioning, etc. To estimate the water demands of these common applications, reference can be made to the summary at Table 4.1.

4.2.2 Supply Sources

A wide range of wastewater discharges can be taken as potential sources of reclaimed water. In making a choice, reliability of the discharge quantities is one of the key factors to be considered. For example, rainwater does not have a regular pattern of occurrence and is therefore often regarded as a supplementary source of the reclaimed water supply. Industrial discharges may also vary significantly in quantity depending on the plant operation mode.

The wastewater quantities discharged from residential buildings, hotels and restaurants are relatively more stable. Their estimates can be based on the water consumption records with an adjustment factor of 0.8–0.9 to account for water consumed but not discharged. Another adjustment factor of 0.67–0.91 has to be applied to the maximum daily water consumption figures to account for the seasonal variation. A breakdown of the various types of discharge can be worked out by

Table 4.1 Water demand estimation factors

Reclaimed water uses	Demand estimation factors
(a) Toilet flushing	• Persons using toilets per day • No. of toilet flushing per person per day • Water quantity of each flushing (6–12 L) • Alternatively, based on percentage of total fresh water consumption (see Table 4.2.)
(b) Landscape irrigation and street cleansing	• Area to receive spraying • Water jet pressure and flow rate • No. of rounds of spraying per day • Duration of spraying
(c) Car washing	• No. of cars to be washed per day • Water quantity for washing each car (small car 250–400 L, bus 400–600 L)
(d) Air-conditioning	• Evaporation rate (4–8 L/hour) • Leakage and waste (16–18 L/hour) • Loss due to wind (25 L/hour)

Table 4.2 Consumption percentages

Consumption item	Residential building (%)	Hotel (%)	Restaurant (%)
Toilet flushing	21.3–21	10–14	6.7–5
Kitchen	20–19	12.5–14	93.3–95
Bath tub and shower	29.3–32	50–40	—
Hand washing	6.7–6.0	12.5–14	—
Clothe washing	22.7–22	15–18	—
Total	100	100	100

making reference to the consumption percentages as shown in Table 4.2. These figures are suggested by GB 50336-2002, 'Code of design for building reclaimed water system' adopted in China.

A reclaimed water system should have a designed supply capability equivalent to 110–115% of the estimated demand. From Table 4.2, it can be seen that with the wastewater collected from bathing, hand washing and clothe washing, excluding that from the kitchens, the combined sources should be able to produce reclaimed water at least adequate to meet the toilet flushing demand, with substantial surplus quantities for other uses. If the discharges from toilets and kitchens are included also, the quantity of reclaimed water produced should be even more, only that the water treatment process will then be more complicated.

4.2.3 Storage and Backup Supply Facilities

As one can expect, there are always discrepancies between the supply and demand patterns of a reclaimed water system. There is also a time lag between the wastewater discharge and the production of reclaimed water in the treatment (see Section 4.3.3) and delivery processes. A storage tank should therefore be provided wherever the inflow rate falls short of the outflow rate at times, and the tank capacity should be designed to avoid emptying during normal operation. Excessive quantities are usually discharged into the sewage drains. If the quality is good enough, the excess flow can be directed to local water catchments to replenish the raw water resources.

In case the supply rate and the tank storage cannot meet the demand, a backup fresh water supply arrangement is needed at the main supply tank to ensure uninterrupted supply to the end-users. This fresh water supply must be controlled in such a way that it will be turned on only when the tank storage falls below a preset low level and will be turned off once the storage returns above that level. The fresh water feeding point should be kept well above the tank overflow level, with a recommended air gap from the water surface of at least 2.5 times the supply pipe diameter (GB 50336-2002). This is to prevent contamination of the fresh water source due to the reclaimed water being suck back into the fresh water system when the supply pressure drops.

4.3 QUALITY CONSIDERATIONS

4.3.1 Quality Requirement

There are different water quality standards for different purposes of use, and for multiple purposes the highest quality standard for the intended applications should be adopted. To meet the standard, the quality of wastewater serving as the source has to be identified and the water has to undergo effective water treatment processes necessary to remove the pollutants as required. While the resulting purified portion is taken for use as reclaimed water, the rejected portion needs to be properly disposed of at the same time.

Although reclaimed water is seldom treated up to the standard for potable use, its quality has to be good enough for the intended purpose, without creating health hazards, objectionable environmental impacts and supply system operational problems. First of all, the water must be aesthetically acceptable to the end-users, particularly in appearance (no foaming or staining), colour and odour (near treatment plant, in toilets, etc.). The bacteria count in the water has to be suppressed to a minimum to guard against health hazards arising from pathogens being carried around in water droplets that may come in contact with people. Toxic substances should also be largely removed to avoid possible contamination of soil, ground water and vegetation for food. Suspended particles should not be allowed to cumulate in the water delivery systems so as to minimise operational problems like encrustation, corrosion and blockage.

Many countries have their own water quality standards to address specific local concerns. There may also be new discoveries of undesirable long-term effects of certain trace chemicals in the wastewater, e.g. antibiotics, hormones, etc., calling for the need to set new standards. In China, the following national standards have been issued pertaining to the reuse of urban recycling water:

- GB/T 18919-2002 on classified standard
- GB/T 18920-2002 on water quality standards for urban miscellaneous water consumption (abstracted in Table 4.3 for reference)
- GB/T 18921-2002 on water quality standard for scenic environment use
- GB/T 19772-2005 on water quality standard for groundwater recharge
- GB/T 19923-2005 on water quality standard for industrial uses.

4.3.2 Quality of Source Water

It is inevitable for mixed wastewater to have varying quality, but the reception storage can provide a balancing effect. Variations within an expected range should not cause major problems, but if there is a sudden surge of pollutants there would be operational difficulties to meet the target quality. Actual sampling and analysis of the source water are important for the design of a reclaimed water system, and also for later monitoring at the operation stage.

In the absence of real data, reference can be made to relevant effluent statistics and standards. Extracted from GB 50336-2002, adopted in China, Table 4.4

Table 4.3 Reclaimed water quality standards for urban miscellaneous uses

Item	Description	Toilet flushing	Road cleansing fire fighting	Urban landscaping	Car washing	Construction use
1.	pH			6.0–9.0		
2.	Colour ≦			30		
3.	Odour			No unpleasant odour		
4.	Turbidity ≦	5	10	10	5	20
5.	Total dissolved Solids (mg/L) ≦	1500	1500	1000	1000	—
6.	BOD_5 (mg/L) ≦	10	15	20	10	15
7.	NH_3-N (mg/L) ≦	10	10	20	10	20
8.	LAS (mg/L) ≦	1.0	1.0	1.0	0.5	1.0
9.	Fe (mg/L) ≦	0.3	—	—	0.3	—
10.	Mn (mg/L) ≦	0.1	—	—	0.1	—
11.	DO (mg/L) ≧			1.0		
12.	Cl_2 (mg/L) ≦			After 30 min. contact ≧ 1.0, at pipe end ≧ 0.2		
13.	Total coliform			3 No./L		

provides some residential wastewater quality figures for reference. For comparison, Table 4.5 sets out the corresponding parameters taken from the secondary sewage treatment standard as recommended in CJ 3025-1993, 'Wastewater and sludge disposal standard for municipal wastewater treatment plants'.

From Tables 4.4 and 4.5, it appears that the quality of secondary sewage treatment effluent is even better than that of the wastewater directly discharged from hand washing basins. However, care should be taken if the sewage treatment plant receives wastewater from hospitals or factories where the discharges may carry pathogens or toxic substances. It is always prudent to select the appropriate wastewater sources instead of relying too much on the treatment processes to minimise undesirable quality elements. There are always practical water quality analysis problems, and thus the unknown risks of harmful substances remaining in the water even after treatment.

Restricting the applications is another way of lowering such risks. The US EPA (1992) has suggested some simple guidelines on the use of reclaimed water, as summarised in Table 4.6.

4.3.3 Water Reclamation Treatment Technologies

Water reclamation treatment technologies are similar to those commonly adopted in sewage treatment, including physical, chemical and biological processes. In general, there are pre-treatment, main treatment and post-treatment stages. The objective is to treat the source water up to the standard suitable for the intended use.

Table 4.4 Reference wastewater quality data in residential areas

	Residential building			Hotel			Restaurant		
Effluent source	BOD_5	COD_{cr}	SS	BOD_5	COD_{cr}	SS	BOD_5	COD_{cr}	SS
Toilet flushing	300–400	800–1100	350–450	250–300	700–1000	300–400	260–340	350–450	260–340
Kitchen sink	500–650	900–1200	220–280	400–550	800–1100	180–220	500–600	900–1100	250–280
Bath tub and shower	50–60	120–135	40–60	40–50	100–110	30–50	—	—	—
Hand washing	60–70	90–120	100–150	50–60	80–100	80–100	—	—	—
Clothe washing	220–250	310–390	60–70	180–220	270–330	50–60	—	—	—
Combined	230–300	455–600	155–180	140–175	295–380	95–120	490–590	890–1075	255–285

Table 4.5 Part of effluent standard for secondary sewage treatment

Parameter	BOD_5	COD_{cr}	SS
Effluent standard	<30	<120	<30

At the pre-treatment stage, the process is to trap and remove suspended particles by physical methods, such as screening, oil trap, etc., and to stabilise the fluctuating quantities and qualities of the incoming water by balancing storage.

At the main treatment stage, the process is to remove the pollutants up to the standard required. Depending on the nature and concentration of the pollutants, a combination of various treatment processes are normally adopted. Common processes include flocculation, sedimentation, dissolved air floatation, biological treatment using membrane bioreactors and sand or membrane filtration.

At the post-treatment stage, the process is to remove bacteria and viruses by disinfection. The treated water then goes into the storage tank for supply.

Table 4.6 Summary of U.S. EPA suggested guidelines for water reuse

Level of treatment	Types of reuse	Reclaimed water quality	Reclaimed water monitoring	Setback distances
1. Disinfected tertiary (filtration of secondary effluent)	• Urban reuse (see Table 4.1) • Food crop irrigation • Recreational impoundments	pH = 6–9 $BOD_5 \leqq 10$ mg/L Turb. $\leqq$ 2NTU E. coli = none Res. $Cl_2 \geqq 1$ mg/L	Weekly Weekly Constantly Daily Constantly	15 m from potable water supply wells. 150 m if impoundment bottom is not sealed
2. Disinfected secondary	Restricted access area irrigation Food crop irrigation (commercially processed) Nonfood crop irrigation Landscape impoundments (restricted access) Construction Wetlands habitat	pH = 6–9 $BOD_5 = 30$ mg/L TSS = 30 mg/L E. coli = 200/100 mL Res. $Cl_2 \geqq 1$ mg/L	Weekly Weekly Daily Daily Constantly	30 m to areas accessible to the public (if spray irrigation) 90 m to potable water supply well.

In selecting the treatment processes, it is advisable to make reference to the performance of similar existing reclaimed water projects. It is also important to conduct a small-scale pilot test to ascertain their effectiveness under the conditions of the project. In Hong Kong, the Government has conducted a pilot water reclamation scheme at Ngong Ping, a tourist area on the mountaintop, collecting the local sewage and wash water for tertiary treatment (see Chapter 5). The effluent is used for toilet flushing and controlled landscape irrigation. It is also used to rear aquarium fish in an exhibition area for publicity purpose (DSD 2006).

4.4 SUPPLY SYSTEM SCALE OPTIONS

4.4.1 General

Reclaimed water supply systems can be broadly classified by scale as municipal, district and building lot. These systems all consist of water collection, treatment, storage and distribution facilities. There will be dual systems for distribution, and may be for wastewater collection also.

Because of the difference in water quality, the reclaimed water system must be entirely separated from the potable water supply system. At the individual receiving buildings, precaution measures have to be taken to prevent cross contamination and misuse. The pipes for reclaimed water supply should have a special colour code. Water tanks and other facilities of the reclaimed water system should be clearly labelled, particularly at the draw-off points.

In planning a reclaimed water project, an economic appraisal of all viable options is necessary. The analysis should include not only the cost of providing reclaimed water as compared to the cost of potable water saved, but also other cost implications and economic factors, such as the cost of exploiting new water resources to cope with extreme drought conditions, savings due to reduced sewage treatment volume, economic benefits of having more reliable water supply for landscape irrigation and air-conditioning, etc. Sustainable financial resources to implement the project and subsequently to operate and maintain the system have to be identified and secured at an early stage. Availability of space for the treatment and pumping facilities is also an important factor of consideration.

4.4.2 Municipal Scale

On the municipal scale, the reclaimed water treatment plant is normally attached to a sewage treatment works, taking the secondarily treated effluent as the source water. The reclaimed water is supplied to individual buildings via a distribution network with pumping, for non-potable uses. The water can also be tapped off along the supply route for street cleansing, drain flushing and landscape irrigation in public areas.

Advantage can be taken to maximise the land use of the same site of the sewage treatment plant, with almost no additional environmental impact. The unit production cost can be lower than that of the other options due to economy of scale. Compared to the discharges from individual sources, the secondarily treated sewage effluent is more stable and reliable in both quality and quantity because of the large quantity collected for treatment. The distribution costs would depend on the remoteness of the sewage treatment plant and the extent of the reclaimed water supply network.

4.4.3 District Scale

On the district scale, if there is a sewage treatment works, the treated effluent can be taken as the source of reclaimed water, similar to that on the municipal scale. If primary sewage treatment is provided, upgrading of the sewage treatment process is necessary for the production of reclaimed water.

If there is no sewage treatment works in the district, the source water may be taken from the rainwater drainage system, which also receives wash water discharged from the buildings of the district, but not the foul water from toilets and the wash water from kitchens to avoid water treatment complications. During dry weather periods, the wastewater, together with any surface runoff collected, will be largely utilised as the source for reclaimed water. During rainy periods, the reclaimed water treatment system will take the quantity as much as required up to its designed capacity. The excess of the collected flow will be directed to the storm water drainage system downstream. As such, the source water quality is expected to be fluctuating as the rainwater input varies. The reclaimed water treatment processes have to be adjusted accordingly.

After treatment, the reclaimed water is distributed through a separate supply system to individual buildings for non-potable use, while the concentrate rejected from the treatment process is discharged into the public sewer system for disposal. By recycled use of part of the wastewater, the sewage treatment works will have less loading of raw sewage in volume, but more concentrated in pollutants.

The reclaimed water treatment plant can be suitably located to minimise the pumping costs. The wastewater from high-level areas can be collected, treated and distributed to lower areas by gravity.

4.4.4 Building Lot Scale

On the scale of a single building lot, the sources of reclaimed water can be limited to the discharges from bathtubs or showers, hand wash basins, washing machines and rainwater only. Recycled use of discharges from toilets and kitchens needs special consideration of the environmental impacts, as the biological process to treat the heavy loading of organic pollutions may generate unpleasant odour. After treatment, the reclaimed water is distributed via a separate supply system for non-potable use, while the rejected concentrate can be discharged into the public sewer system for disposal.

The unit production cost is expected to be higher than that in the other two options. However, implementation of the project can be independent of the systems in the other building lots around. In fact, in a densely populated city like Hong Kong, it is not uncommon to find a housing estate accommodating over 10,000 people, with wash water discharges of around 1000 m^3/day. Reuse of the wash water for non-potable use within the estate could be economically attractive, if the full cost of potable water supply (without subsidy from the government) is taken into account. It would be even more attractive if the sewage charges are not applied to the recycled quantities.

With the advancement of water treatment technologies and the reduction in cost of many water treatment products, such as membranes, it may be viable to develop cost-effective reclaimed water systems in a creative manner. For example, the wastewater discharged at a great height of a high-rise building may be collected mid-way down the building, e.g. at the firebreak level, where pre-treatment processes can take place. The head difference between the pre-treatment storage tank at that level and the treatment facilities at the ground level can be utilised to effect membrane filtration, thus substantially saving the operation cost. A pilot plant study may convert such an idea into practice.

The use of reclaimed water can definitely achieve a number of benefits in respect of environmental and water resources management. Many projects have been successfully implemented and put in use for decades. However, there are still scope for further development in technologies and creative designs to provide reclaimed water in a more cost-effective way for wider application.

4.5 REFERENCES

CJ 3025-1993. Wastewater and Sludge Disposal Standard for Municipal Wastewater Treatment Plants. The Ministry of Construction of the People's Republic of China.

Drainage Services Department (DSD). 2006, Use of Reclaimed Water. Government of HKSAR (www.dsd.gov.hk/sewerage/use_of_reclaimed_water/index.htm).

GB 50336-2002. Code of Design for Building Reclaimed Water System. The General Administration of Quality Supervision, Inspection and Quarantine of the People's Republic of China.

GB/T 18920-2002. The Reuse of Urban Recycling Water – Water Quality Standard for Urban Miscellaneous Water Consumption. The General Administration of Quality Supervision, Inspection and Quarantine of the People's Republic of China.

Ku, C. C., Damien, Fung, C. K. and Chan, T. H. 2003, Utilisation of seawater as part of Total Water Management in Hong Kong. Proceedings of the Second Symposium on Sustainable Development of Guangdong, Hong Kong and Macau – Strategic Partnership in the Pearl River Delta.

US Environmenatl Protection Agency (EPA). 1992, Manual – Guidelines for Water Reuse. EPA/625/R-92/004, U.S. Environmental Protection Agency and U.S. Agency for International Development, Washington DC.

Water Supplies Department (WSD). 2005, 2005 Key Facts. Government of HKSAR (www.wsd.gov.hk/en/html/info_publ/stat/index.htm).

5

Cost comparison of seawater, raw fresh water and reclaimed water for toilet flushing

5.1 BACKGROUND INFORMATION

As mentioned in the previous chapters, fresh water in Hong Kong is obtained from the local rainfall and raw water purchased from Dongjiang (East River) of China. The total fresh water consumption in 2003 was 960 million m^3 per year, while 744 million m^3 per year was transported from Dongjiang (WSD 2003b). Except for a small number of villages scattered in the remotest areas, over 99.9% of Hong Kong's population receives piped fresh water supply of purity that is substantially better than the WHO standard. In the seawater supply system, seawater is directly extracted from the sea and supplied to about 80% of the Hong Kong population for toilet flushing with the annual consumption amounting to 235 million m^3 in 2003. Seawater supply is a free resource and there is no restriction on the amount used. The main advantages of using seawater are its abundant supply and contribution in fresh water conservation.

The utilisation of both fresh water and seawater resources, from their abstraction to their final disposal, would have different degrees of impact on the environment. From a sustainable perspective, rational utilisation of water resources, both for the existing ones and the new resources to be explored, should be considered and is

S. L. Tang, Derek Yue and Damien Ku. ISBN: 9781843391326. Published by IWA Publishing, London, UK.

important for the long-term development of Hong Kong. The Hong Kong Government, therefore, pledged in the 2003 Policy Agenda that a total water management (TWM) programme would be implemented as a major initiative to enhance water conservation and water resource protection. The TWM involves optimal utilisation of water resources in the whole water cycle and adopts a two-pronged approach. On one hand, it encourages the public to reduce consumption through education and publicity programmes, minimise non-effective uses such as leakages and the use of alternative resources other than fresh water for toilet flushing through the implementation of capital projects. On the other hand, it encourages the exploration of new water resources in order to increase the flexibility and versatility in securing adequate and reliable water supply.

As an integral part of the TWM programme, a pilot scheme for reclaimed water (treated effluent from sewage treatment plant) reuse has been launched, as was described in Chapter 4. A tertiary sewage treatment plant at Ngong Ping, a tourist attraction in Lantau Island of Hong Kong, had been constructed and was commissioned in 2005 to supply reclaimed water for toilet flushing at the cable car terminal and public toilets. This scheme can reduce the demand of fresh water and reduce effluent from the sewage treatment plant to be discharged into the environment (ACQWS 2003). According to the annual report of WSD (2003d), the Peak, the eastern part of the Southern District of Hong Kong Island, Yuen Long, Tin Shui Wai, Sheung Shui, Fanling, Sai Kung and the outlying islands are not at present equipped with the seawater flushing system. The absence of it is mainly due to the high installation cost as these areas are not close to the seafront, or where the population is scattered. Since Ngong Ping is located at one of the outlying islands, the experience obtained from such a pilot scheme can realistically verify the feasibility of reclaimed water supply, as an alternative water source, for toilet flushing in these non-seawater supply zones.

Seawater is currently supplied to about 80% of the population of Hong Kong for toilet flushing. The seawater supply zone however only covers the urban areas on both sides of the Victoria Harbour and some new towns in the New Territories that are close to the seafront. Potable water is therefore used in the remaining areas (i.e. non-seawater supply zones), where the population is scattered and sparse, for both domestic consumption and toilet flushing. In order to conserve the valuable fresh water resource, 25% of the population (see explanation below for 25%) in each of the four districts—Southern District (SD) of Hong Kong Island, Sai Kung (SK), Northern New Territories (NNT) including Sheung Shui and Fanling and Northwest New Territories (NWNT) including Tin Shui Wai and Yuen Long—is assumed to have a separate toilet flushing supply system installed in this study. These selected four districts are highlighted in Figure 5.1. The reasons for such selection are that the population in these districts is denser than other non-seawater supply areas in the Territory and that a number of major secondary sewage treatment works (STW) have already been installed and operated by Drainage Services Department (DSD) in the districts as shown in the same figure. As limited by the treatment capacity of the existing STWs, the reclaimed water can only supply to 25% of the population in the areas under study (see Table 5.7).

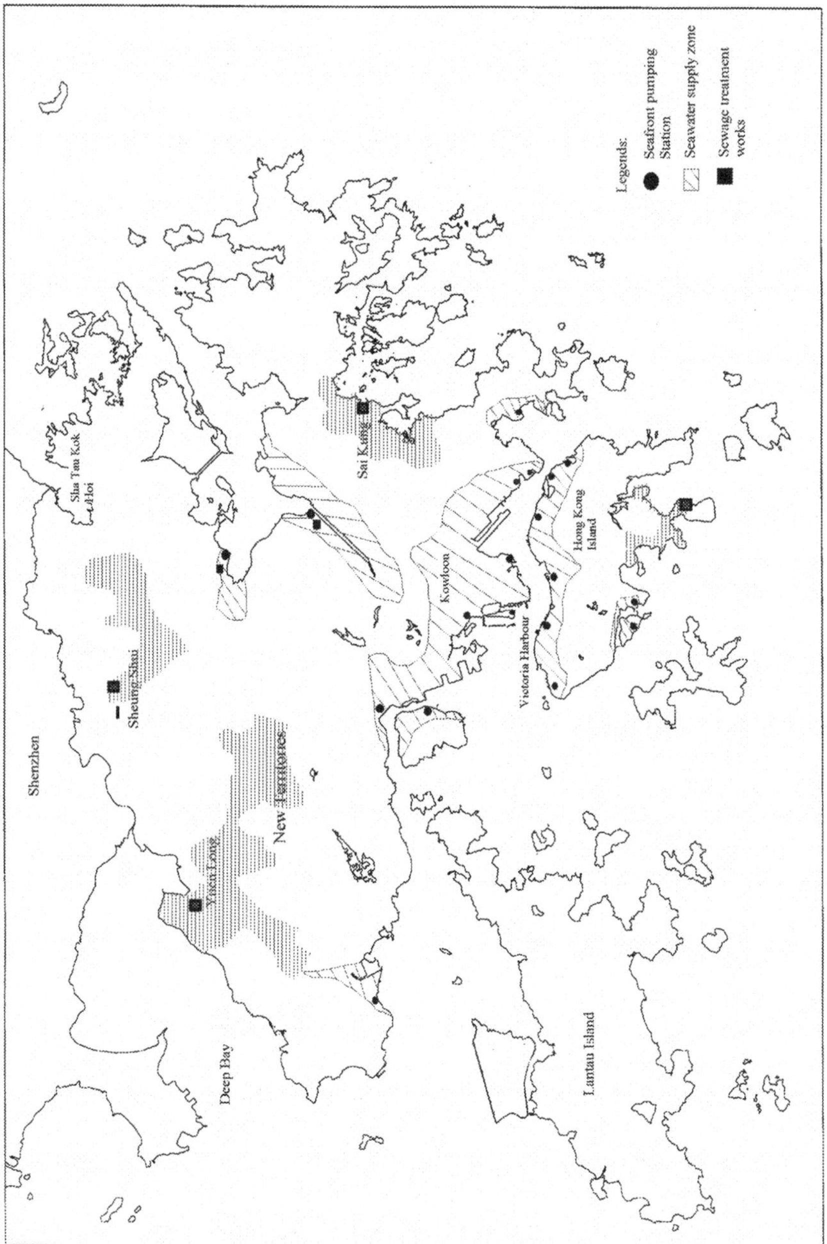

Figure 5.1 Proposed new flushing water supply zones. Fully shaded areas = Proposed flushing water supply zones by sea water (Case 1), raw water (Case 2) and reclaimed water (Case 3).

Seawater, raw fresh water from Guangdong of China, and reclaimed water from STWs are assumed to be the three water resources for the separate flushing system in these districts. The objective of this chapter is to carry out a cost comparison among the following cases and find out which one has the best engineering economy. The chapter is written based on the research of (Tang, S. L., Yue, Derek, P. T. and Li, X. Z.).

Case 1: The toilet flushing system using seawater in SD, SK, NNT and NWNT.
Case 2: The toilet flushing system using raw water in SD, SK, NNT and NWNT.
Case 3: The toilet flushing system using reclaimed water (treated effluent) in SD, SK, NNT and NWNT.

5.2 FLUSHING WATER RESOURCES

5.2.1 Seawater

Since 1950s, the Water Supplies Department (WSD) has supplied seawater, primarily for toilet flushing, in government and government-aided high-density development buildings at an early stage, and subsequently supplied throughout the urban areas and the new towns. About 80% of the population is currently supplied with seawater for toilet flushing (WSD 2003b). The extensive use of seawater has helped reduce the demand on fresh water for toilet flushing. Recently in Hong Kong, an average of 660,000 m^3/day seawater was supplied for flushing purpose, conserving an equivalent amount of potable water.

The total capital cost of the seawater distribution system in Hong Kong is HK$3738.9 million valued in 2002 (WSD 2003c), which is shown in Table 5.3. After applying the consumer price index (CPI) adjustment (CSD 2003), this value becomes HK$3944 at year 2000 money value.

Further details for seawater resource can be found in the previous chapters.

5.2.2 Raw Fresh Water

The supply of raw fresh water from Guangdong, a province in Southern China, is now the major source of raw fresh water supply for Hong Kong. It has a history dated back to 1960s when a scheme was first formulated for receiving a supply of 22.7 million m^3 a year. The second, third and fourth agreements were reached subsequently in 1964, 1978 and 1987, respectively, with the Guangdong Authorities whereby Hong Kong could purchase from 68 million m^3 in 1964 to 690 million m^3 in 1995. Under the supply agreement of 1989 and then the supplementary supply agreement of 1998, the maximum annual supply quantity of raw water for 2001 was 790 million m^3 with an annual increase of 10 million m^3 up to 820 million m^3 in 2004 (WSD 2003a,d). Discussions continued to be held with the Guangdong Authorities on matters of price, quality assurance and flexibility of supply. The rest of the raw water in Hong Kong is rainfall gathered from local catchments.

Water extracted from Dongjiang (East River), at a point some 83 km north of Hong Kong, is pumped over a series of dams built across the Shima River, one of its tributaries. It is eventually discharged into the Shenzhen Reservoir before being fed by pipelines to Shenzhen Special Economic Zone and is then brought across the border to Hong Kong. The Dongjiang water supplied to Hong Kong is expected to conform to the *Environmental Quality Standard of Grade 2 Surface Water* published by the People's Republic of China. The Guangdong Authorities have also carried out the following measures in recent years so as to assure the quality of Dongjiang raw water transferred to Hong Kong:

(a) STW have been built along the areas near to the transfer aqueduct and its drainage basin to treat sewage before discharging into the aqueduct.
(b) Legislation has been enacted to prohibit establishment of factories that may pollute the water transferred and to remove similar existing factories.
(c) The intake of Dongjiang water has been moved upstream in 1998 where the water quality is better.

In addition to these measures, the commissioning of the biological nitrification plant at the Shenzhen Reservoir since early 1999 to eliminate ammonia nitrogen content in the raw water and the project of constructing a closed aqueduct from the Dongjiang intake to Shenzhen Reservoir completed in 2003 to avoid contamination from local industries have also greatly improved the quality of the Dongjiang water as raw fresh water supply to Hong Kong. The raw water, after further purification by water treatment plants in Hong Kong, is suitable and safe for drinking.

If the raw water is to be used for flushing, full treatment of it at fresh water treatment plants in Hong Kong is not necessary. Partial treatment at a much lower cost is adequate for such a purpose. Such partial treatment will include chlorination and dyeing processes, similar to the reclaimed water treatment (see Tables 5.1 and 5.2). As agreed in the latest long-term water supply agreement with Guangdong Authorities, the unit raw water (untreated) price was capped at HK\$3.085 per m^3 since 1999.

Table 5.1 Capital cost of tertiary treatment and effluent reuse treatment

		Design capacity (Mm^3/year)	Construction cost (HK\$ million)	Year basis	CPI	Real cost year 2000 based (HK\$ million)
Tertiary treatment	SBR tanks	1.0788	33	02/03	94.8	34.8101
	Filter and UV disinfection	1.0788	13	02/03	94.8	13.7131
Effluent reuse treatment	Chlorination and colour dye systems	0.3285	5.5	02/03	94.8	5.8017

Mm^3 = million cubic metres.

Table 5.2 E&M cost of tertiary treatment and effluent reuse treatment

		Design capacity (Mm^3/year)	Real total capital cost 2000 basis (HK$ million)	Real E&M cost year 2000 based (HK$ million)
Tertiary treatment	SBR tanks	1.0788	34.8101	13.924
	Filter and UV disinfection	1.0788	13.7131	5.4852
Effluent Reuse treatment	Chlorination and colour dye systems	0.3285	5.8017	2.3207

5.2.3 Reclaimed Water from Local STWs

As mentioned in Section 5.1, a pilot scheme for treated effluent reuse has been launched. A tertiary sewage treatment plant at Ngong Ping has been constructed and was commissioned in 2005 to supply reclaimed water (treated sewage effluent) for toilet flushing. The sewage treatment processes adopted in Ngong Ping STW include screening, grease removal, grit removal, sequencing batch reactor (SBR), tertiary filtration and ultraviolet (UV) disinfection. The SBR is a biological treatment process, which utilises suspended growth of microorganisms to remove organic pollutants and nutrients from the wastewater. As the sewage has to undergo biological nitrification and de-nitrification in the SBR, the ammonia nitrogen can be significantly reduced. The SBR also has the benefits of small size and high capability in handling high fluctuation of sewage flows. The costs of the treatment processes adopted in this project are to be based for the economy comparison. These cost data are shown in Table 5.1.

The cost data in Table 5.1 are originally based on 2002–2003 money value. By using CPI published by the Census and Statistics Department of Hong Kong (CSD 2003), the costs are converted to the money value of year 2000. This is to ensure that the net present value (NPV) calculated is real without having to consider inflation adjustments in the calculation process because all costs will be year 2000 based (Tang 2003; Chapter 4).

The costs of the tertiary sewage treatment and effluent reuse treatment shown in Table 5.1 include both civil engineering works and electrical and mechanical (E&M) equipment. According to the chief design engineer of the Ngong Ping STW, the capital cost of E&M is about 40% of the construction capital cost. Table 5.2 shows the total capital costs and the E&M costs.

As mentioned earlier, this capital costs of tertiary treatment and effluent reuse treatment are obtained from the Ngong Ping STW project. However, this STW is only designed for treating 1.0788 Mm^3 sewage per year and is a very small-scale plant compared to the existing STW in the four selected districts under this study. Hence, scale-effect factors should be applied to adjust the capital costs and the operation and administration expenses (OAE; to be discussed in due course) so that the results of the engineering cost comparison will be more realistic. The values of the scale-factors are assumed to be 0.8 for the capital costs and 0.9 for the OAE. Although these are subjective values, the authors think that they are based on common sense and are, in the opinion of the authors, reasonable assumptions.

In this study, because supplying treated effluent for toilet flushing from those existing secondary STWs in the selected districts is assumed, the SBR tertiary treatment and effluent reuse treatment processes are assumed to be not newly built but added to the existing secondary treatment processes. The SBR tertiary treatment is to provide better nitrification and de-nitrification for ammonia-nitrogen removal, while the effluent reuse treatment is to provide further treatment (chlorine dosing and colour dye dosing processes) after the effluent is tertiarily treated. According to the estimates made by DSD of Hong Kong, the OAE of the sewage treatment including coarse screening, grease separate, grit classfier, SBR, dual media granular filter, UV disinfection, sludge digestion and dewater facilities for this project is about HK\$5 to HK\$6 per m^3. The OAE of the effluent reuse treatment is about HK\$2/m^3 (information from Louis Chan of Ove Arup & Partners Hong Kong Limited). In this study, since the SBR, dual media granular filter and UV disinfection process are assumed to be added to the existing secondary STW for tertiary upgrading, and since no other capital cost data are available in Hong Kong for effluent reuse treatment except the Ngong Ping STW, the capital cost of the extra treatment facilities added to the existing STW is assumed to be 85% of that of the Ngong Ping STW. The authors think that such an assumption is not unreasonable, given the situation that there has been no previous experience (data) in this respect. Similarly, the OAE spent on these processes should be much less than that of the Ngong Ping STW as the expenses will only be the additional OAE to the existing secondary STW. Hence, in the analysis given later, the OAE of these additional treatment processes is reasonably assumed to be HK\$2.5/m^3. The OAE of effluent reuse treatment (chlorination and dyeing) is unchanged and maintained at HK\$2/m^3 because it is new and not an additional treatment.

5.3 BASIC COST ASSUMPTIONS AND DATA FOR COST COMPARISON OF USING THE THREE FLUSING WATER RESOURCES

5.3.1 General

In this analysis, the time duration (or time horizon) for comparison are the same for all the cases. A time horizon of 60 years is assumed, which is the least common multiple of three important renewal years of different categories of facilities: 15 years for E&M equipment, 20 years for seawater mains and 30 years for fresh water mains, plant and machinery and meters. These assumptions are made by referring to WSD's operating accounts related to the depreciation of assets (WSD 2003c).

It is assumed that the water demand and supply is constant throughout the running period (60 years). This is merely an assumption of course. In fact, the demand of water will increase due to the growth of population, economic development and living standard, and the amount of water supply will increase correspondingly to meet the increasing demand. For this reason, the capital cost and running cost will therefore change with time. However, because this analysis is only for the purpose of engineering cost comparison of the said cases instead of determining an exact absolute real NPV, the consideration for continuous growth of the demand and supply will not be incorporated in the analysis.

The NPV method is used for carrying out the comparison. An introduction of NPV has been given in Section 3.4. The NPV determined from the calculation will be used to directly compare the engineering costs of the cases. The following equation is employed to obtain the NPV (Tang 2003):

$$\mathrm{NPV} = F_0 + \frac{F_1}{(1+i)^1} + \frac{F_2}{(1+i)^2} + \cdots + \frac{F_n}{(1+i)^n},$$

where F_n = net cash flow or 'income—expenditure' in Year n. Expenditure includes running cost for each year ($n = 1, 2, \ldots, 60$), the capital cost when $n = 0$ and the renewal cost when $n = 30$ for the fresh water, raw water and reclaimed water supply systems, and the renewal cost when $n = 20$ and 40 for the seawater supply system. E&M items are renewed when $n = 15$. The incomes for the cases are not considered and are assumed to be the same in all cases because they are practically so. If they are not considered, their values are set as 0 when the NPV formula is applied, and so the NPV calculated for each of the cases is in fact the present value of total cost. This will adequately serve our purpose because our objective is to compare the present values of total costs of the cases.

5.3.2 Capital Costs and Recurrent Costs of Existing Facilities

The capital costs, staff costs and OAE of the existing facilities (WSD 2003c) are shown in Tables 5.3–5.5. All costs have been converted, as explained before, to year 2000 base year prices.

In Tables 5.4 and 5.5, it can be observed that the staff cost of HK$1501 million per year and the OAE of HK$1297 million per year are the weighted average figures of the past 10 years.

5.3.3 Capital Cost of the Extra Flushing Water Supply Pipelines

In order to supply flushing water to 25% of the population of the proposed four districts (SD, SK, NNT and NWNT, as described in Section 5.1), a new flushing pipeline system must be provided. The total population in these districts is 21.2%

Table 5.3 Capital cost of existing facilities

Fixed assets	Cost at 2002 (HK$ million)	CPI at 2002	Real cost 2000 basis (HK$ million)
Dams and other works	120.0	94.8	126.6
Buildings, filters, mains, etc.	24,971.1	94.8	26,340.8
Plant and machinery	177.6	94.8	187.3
Salt water flushing	3,738.9	94.8	3,944.0
Tai Lam Chung Reservoir	132.8	94.8	140.1
Shek Pik Reservoir	285.9	94.8	301.6
Plover Cove Reservoir	702.0	94.8	740.5
High Island Reservoir	1,661.2	94.8	1,752.3
Meters	320.3	94.8	337.9
Motor vehicles	50.3	94.8	53.1
Assets under construction	6,279.1	94.8	6,623.5
Total	38,439.2	94.8	40,547.7

Source: WSD 2003c.

of that of the whole territory. Since flushing water is to be supplied to only 25% of this population, the theoretical additional capital cost of the flushing pipeline system in the four selected districts is

$$\$3944 \times 21.2\% \times 25\% = \$209 \text{ million},$$

where $3944 million is the capital cost of the existing flushing system at year 2000 money value (see Table 5.3). However, since the total area of these four districts is

Table 5.4 Staff cost of the existing system

Year	CPI	Staff cost (HK$ million)	Real staff cost at 2000 basis (HK$ million)	Fresh water consumption (Mm^3)	Product
1993/94	78.3	1,013	1,293.7420	915	1,183,773.9464
1994/95	85.2	1,118	1,312.2066	923	1,211,166.6667
1995/96	92.9	1,275	1,372.4435	919	1,261,275.5651
1996/97	98.8	1,396	1,412.9555	928	1,311,222.6721
1997/98	104.5	1,474	1,410.5263	913	1,287,810.5263
1998/99	107.5	1,603	1,491.1628	916	1,365,905.1163
1999/2000	103.2	1,679	1,626.9380	911	1,482,140.5039
2000/01	99.4	1,734	1,744.4668	924	1,611,887.3239
2001/02	97.8	1,747	1,786.2986	940	1,679,120.6544
2002/03	94.8	1,473	1,553.7975	949	1,474,553.797
		Total	15,004.53747	9,238	13,868,856.7725
				Weighted mean	1,501.2835

Source: WSD 2003d.

Table 5.5 OAE of the existing system

Year	CPI	OAE (HK$ million)	Real OAE at 2000 basis (HK$ million)	Fresh water consumption (Mm^3)	Product
1993/94	78.3	832	1,062.5798	915	972,260.5364
1994/95	85.2	978	1,147.8873	923	1,059,500.0000
1995/96	92.9	1,177	1,266.9537	919	1,164,330.4629
1996/97	98.8	1,147	1,160.9312	928	1,077,344.1296
1997/98	104.5	1,247	1,193.3014	913	1,089,484.2105
1998/99	107.5	1,303	1,212.0930	916	1,110,277.2093
1999/2000	103.2	1,332	1,290.6977	911	1,175,825.5814
2000/01	99.4	1,330	1,338.0282	924	1,236,338.0282
2001/02	97.8	1,485	1,518.4049	940	1,427,300.6135
2002/03	94.8	1,674	1,765.8228	949	1,675,765.8228
		Total	12,956.7000	9,238	11,988,426.5945
				Weighted mean	1,297.7297

Source: WSD 2003d.

quite large and the population is relatively scattered, the flushing pipeline system should therefore be more expensive. It is assumed that the flushing pipeline system will be twice as expensive as that built in the existing seawater supply zone. The assumption is made based on the following reason: the population in these four districts altogether is about 20% of Hong Kong's total population. The population in the existing seawater supply zone (mainly city areas) is about 80% of the total population, around four times the total population in the four districts. The total area, however, of the four districts and that of the existing seawater supply zone are about the same (see Figure 5.1). Therefore, the population density of the latter is four times that of the former. Figure 5.2A shows that the average area lived by one person in the city areas (existing seawater supply zone) is y^2 m^2, and Figure 5.2B

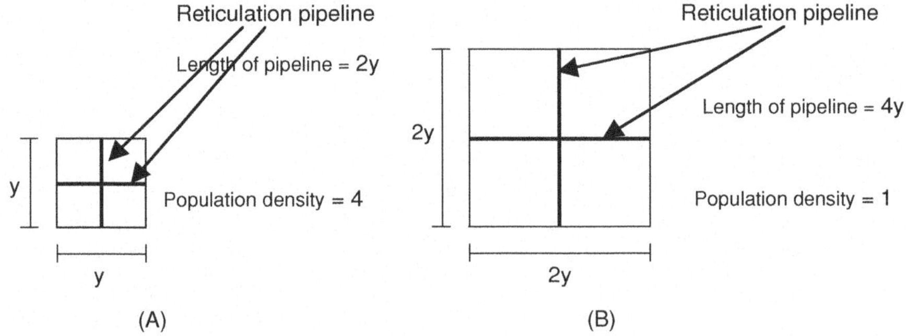

Figure 5.2 Relationship between population density and length of reticulation pipeline system.

Table 5.6 Capital cost of flushing system in each selected district

District of STW	Population ratio	Capital cost of flushing system (HK$ million)
SD	0.198	82.764
SK	0.249	104.082
NNT	0.202	84.436
NWNT	0.351	146.718
Total	1	418

shows that the average area lived by one person in the four districts in question is $4y^2$ m^2. It follows that the total length of the seawater pipeline reticulation system in the four districts should be $\sqrt{4}$ times (or 2 times) of the total length of that for the existing seawater supply zone. Therefore, when the population density of an area is four times denser, the total length of the pipeline reticulation system in that area should be 2 (i.e. $\sqrt{4}$) times longer. Hence, a more accurate estimate of the additional capital cost of the flushing system in these four selected districts is

$$= \$209 \text{ million} \times 2 = \$418 \text{ million.} \tag{5.1}$$

Hence, the capital investment of the whole system in Hong Kong is (see Table 5.3 and Eq. (5.1))

$$\begin{aligned} &= \$126.6 + 26340.8 + 187.3 + (3944 + 418) + 140.1 \\ &\quad +301.6 + 740.5 + 1752.3 + 337.9 + 53.1 + 6623.5 \\ &= \$40,965.7 \text{ million} \end{aligned} \tag{5.2}$$

Therefore, the percentage of increase of capital investment is (see Table 5.3)

$$\begin{aligned} &= \frac{40,965.7}{(40,965.7 - 418)} \times 100\% \\ &= 101.031\%. \end{aligned} \tag{5.3}$$

It can be seen that there is a 1.031% increase in the value of the facilities. This 1.031% increase can be used to calculate the increase in the overall costs of staff and OAE.

By referring to Figure 5.1, the population ratios of SD, SK, NNT and NWNT are listed in Table 5.6, and they are used to determine the capital cost of the flushing water supply pipeline system in each district, which is also shown in Table 5.6.

5.3.4 Quantity of Flushing Water Supply in the Four Districts

According to WSD's annual report (WSD 2003d), the annual fresh water and seawater consumption are 949 and 235 Mm^3, respectively. From past experience, flushing water is about 20% of the total water consumption. As the total population

Table 5.7 Treatment capacity of existing STW in the selected districts

District of STW	Name of STW	Treatment capacity (Mm^3/year)	Population ratio	Flushing water required (Mm^3/year)
SD	Stanley	2.555	0.198	2.4814
SK	Sai Kung	3.285	0.249	3.1267
NNT	Shek Wu Hei	25.55	0.202	2.5364
NWNT	Yuen Long	17.52	0.351	4.4058
Total		48.91	1	12.5504

of the four selected areas is 21.2% of that of the whole territory, the required quantity of flushing water is (see Table 5.7)

$$\begin{aligned} &= (949 + 235) \times 20\% \times 21.2\% \times 25\% \\ &= 12.5504\,\mathrm{Mm^3/year.} \end{aligned} \tag{5.4}$$

The treatment capacities of the existing STW in the selected districts are also shown in Table 5.7.

5.4 COST COMPARISON FOR CASES 1–3

By using the data given in the previous section, the cases under consideration have their each NPV calculated. The details are shown next.

5.4.1 Case 1: The Toilet Flushing System Using Seawater in SD, SK, NNT and NWNT

As the districts of NNT and NWNT are far away from shore, large diameter pipelines are required to be constructed to supply seawater to these two districts for toilet flushing. In addition, export pipelines are also required for discharging effluent from the existing sewage treatment plants in these two districts. Export pipelines are necessary because saline effluent coming out from STW must be discharged back to the sea and not rivers or nearby surroundings, as explained in Section 2.7. As shown in Figure 5.1, the seawater can be abstracted and discharged at Sha Tau Kok Hoi and Deep Bay for NNT and NWNT, respectively, with about 10 km long pipelines. The diameter of import and export pipelines can be estimated by assuming the flow velocity in the pipes to be between 2 and 3 m/second as follows.

In NNT, the required average flow of import pipeline is (see Table 5.7)

$$\begin{aligned} &= \frac{2.5364 \times 10^6}{365 \times 24 \times 3600} \\ &= 0.08042\ \mathrm{m^3/second.} \end{aligned}$$

Therefore, the diameter of import pipeline is

$$= \sqrt{\frac{0.08042/3}{3.1416}} \tag{5.5a}$$

$$= 0.092 \text{ m (say 150 mm diameter).}$$

Similarly in NWNT, the diameter of import pipeline is (see Table 5.7)

$$= \sqrt{\frac{4.4058 \times 10^6/(365 \times 24 \times 3600)/3}{3.1416}} \tag{5.5b}$$

$$= 0.1218 \text{ m (say 150 mm diameter).}$$

For the export pipelines, the size of pipes is capable of discharging the effluent from the STW. Hence in NNT, the required average flow in export pipeline is (see Table 5.7)

$$= \frac{25.55 \times 10^6}{365 \times 24 \times 3600}$$

$$= 0.81 \text{ m}^3/\text{second.}$$

The diameter of export pipeline is

$$= \sqrt{\frac{0.81/3}{3.1416}} \tag{5.6a}$$

$$= 0.293 \text{ m (say 300 mm diameter).}$$

Similarly in NWNT, the diameter of import pipeline is (see Table 5.7)

$$= \sqrt{\frac{17.52 \times 10^6/(365 \times 24 \times 3600)/3}{3.1416}} \tag{5.6b}$$

$$= 0.24 \text{ m (say 300 mm diameter).}$$

According to the research report of (Tang, S. L., Yue, Derek, P. T. and Li, X. Z.) (2003), the construction cost of 150- and 300-mm diameter pipelines are HK$1466.673 and HK$2065.323 per m, respectively, in Hong Kong at year 2000 money value. From the maintenance point of view, a pair of pipelines (i.e. twin pipes) should be constructed instead of a single pipeline. Therefore, the total capital investment of the import pipelines for NNT and NWNT is

$$= \frac{4 \times 10 \times (1466.673 \times 1000)}{10^6} \tag{5.7a}$$

$$= \text{HK\$58.667 million.}$$

The total capital investment of the export pipelines for NNT and NWNT is

$$= \frac{4 \times 10 \times (2065.323 \times 1000)}{10^6} \tag{5.7b}$$

$$= \text{HK\$82.613 million.}$$

The following is a summary of costs for Case 1.

Capital Investment (see Eq. (5.1), (5.7a) and (5.7b)) = Additional flushing system + import and export pipelines = 418 + 58.667 + 82.613	559.28
Total running cost (see Tables 5.4 and 5.5 and Eq. (5.3)) = Staff cost + OAE = (1,501.2835 × 1.031%) + (1297.7297 × 1.031%)	28.858
Renewal cost (seawater, 20 years) of flushing system (see Eq. (5.1), (5.7a) and (5.7b)) = 418 + 58.667 + 82.613	559.28

It should be noted that the construction costs of the import and export pipelines are not considered in the analysis when raw water (Case 2) and reclaimed water (Case 3) are used for toilet flushing because such pipelines are not necessary for these cases, since the impact to the ecology would be minimal when raw water and reclaimed water are used. The result of NPV calculation for Case 1 is shown in Table 5.8. The NPV calculated is –HK$1398 million.

5.4.2 Case 2: The Toilet Flushing System Using Raw Water in SD, SK, NNT and NWNT

As mentioned earlier, the purchase cost of raw water transported from Dongjiang is HK$3.085/m^3 (at 2002 money value). After applying CPI adjustment, the water purchase cost at 2000 money value becomes

$$
\begin{aligned}
&= \frac{3.085 \times 100}{94.8} \\
&= \$3.2542.
\end{aligned} \tag{5.8}
$$

The total water purchase cost for the four selected areas is (see Eq. (5.4))

$$
\begin{aligned}
&= 12.5504 \times 3.2542 \\
&= \$40.8415 \text{ million}.
\end{aligned} \tag{5.9}
$$

Hence, the water purchase cost spent on toilet flushing system in each selected district can be estimated and is shown in Table 5.9.

Effluent reuse treatment (chlorination and colour dyeing processes) must be provided before the raw water can be supplied to users. The following is a summary of costs for Case 2.

Table 5.8 NPV calculation for Case 1

Year	(1) Capital cost	(2) Running cost	(3) Income	(4) = (3) − (2) − (1) Net cash flow	Discount cash flow at $i = 5\%$
0	559			−559	−559.00
1		29	0	−29	−27.62
2		29	0	−29	−26.30
3		29	0	−29	−25.05
4		29	0	−29	−23.86
5		29	0	−29	−22.72
6		29	0	−29	−21.64
7		29	0	−29	−20.61
8		29	0	−29	−19.63
9		29	0	−29	−18.69
10		29	0	−29	−17.80
11		29	0	−29	−16.96
12		29	0	−29	−16.15
13		29	0	−29	−15.38
14		29	0	−29	−14.65
15		29	0	−29	−13.95
16		29	0	−29	−13.29
17		29	0	−29	−12.65
18		29	0	−29	−12.05
19		29	0	−29	−11.48
20	559	29	0	−588	−221.61
21		29	0	−29	−10.41
22		29	0	−29	−9.91
23		29	0	−29	−9.44
24		29	0	−29	−8.99
25		29	0	−29	−8.56
26		29	0	−29	−8.16
27		29	0	−29	−7.77
28		29	0	−29	−7.40
29		29	0	−29	−7.05
30		29	0	−29	−6.71
31		29	0	−29	−6.39
32		29	0	−29	−6.09
33		29	0	−29	−5.80
34		29	0	−29	−5.52
35		29	0	−29	−5.26
36		29	0	−29	−5.01
37		29	0	−29	−4.77
38		29	0	−29	−4.54
39		29	0	−29	−4.33
40	559	29	0	−588	−83.52
41		29	0	−29	−3.92
42		29	0	−29	−3.74
43		29	0	−29	−3.56
44		29	0	−29	−3.39
45		29	0	−29	−3.23
46		29	0	−29	−3.07

(cont.)

Table 5.8 (*Continued*)

Year	(1) Capital cost	(2) Running cost	(3) Income	(4) = (3) − (2) − (1) Net cash flow	Discount cash flow at $i = 5\%$
47		29	0	−29	−2.93
48		29	0	−29	−2.79
49		29	0	−29	−2.66
50		29	0	−29	−2.53
51		29	0	−29	−2.41
52		29	0	−29	−2.29
53		29	0	−29	−2.18
54		29	0	−29	−2.08
55		29	0	−29	−1.98
56		29	0	−29	−1.89
57		29	0	−29	−1.80
58		29	0	−29	−1.71
59		29	0	−29	−1.63
60		29	0	−29	−1.55
					NPV = −1,398.03

Capital investment (see Eqs. (5.1) and (5.13)) = flushing system + chlorination and colour dyeing processes $= 418 + 221.655 \times 0.8$	595.324
Total running cost (see Tables 5.4 and 5.5, Eqs. (5.3), (5.9) and (5.14)) = staff cost + OAE + water purchase cost $= 1501.2835 \times 1.031\% + [1297.7297 \times 1.031\% + 25.1008 \times 0.9] + 40.8415$	92.29
Renewal cost (raw water, 30 years) of flushing system (see Eq. (5.1)) = 418	418
Renewal cost (15 years) of E&M for chlorination and colour dyeing processes (see Eq. (5.15)) $= 88.662 \times 0.8$	70.9296

The result of NPV calculation for Case 2 is shown in Table 5.10. The NPV calculated is −HK$2492 million.

Table 5.9 Water purchase cost in each selected district

District of STW	Population ratio	Water purchase cost (HK$ million)
SD	0.198	8.0751
SK	0.249	10.1749
NNT	0.202	8.2541
NWNT	0.351	14.3374
Total	1	40.8415

Table 5.10 NPV calculation for Case 2

Year	(3) Capital cost	(2) Running cost	(3) Income	(4) = (3) − (2) − (1) Net cash flow	Discount cash flow at $i = 5\%$
0	595			−595	−595.00
1		92	0	−92	−87.62
2		92	0	−92	−83.45
3		92	0	−92	−79.47
4		92	0	−92	−75.69
5		92	0	−92	−72.08
6		92	0	−92	−68.65
7		92	0	−92	−65.38
8		92	0	−92	−62.27
9		92	0	−92	−59.30
10		92	0	−92	−56.48
11		92	0	−92	−53.79
12		92	0	−92	−51.23
13		92	0	−92	−48.79
14		92	0	−92	−46.47
15	71	92	0	−163	−78.41
16		92	0	−92	−42.15
17		92	0	−92	−40.14
18		92	0	−92	−38.23
19		92	0	−92	−36.41
20		92	0	−92	−34.67
21		92	0	−92	−33.02
22		92	0	−92	−31.45
23		92	0	−92	−29.95
24		92	0	−92	−28.53
25		92	0	−92	−27.17
26		92	0	−92	−25.87
27		92	0	−92	−24.64
28		92	0	−92	−23.47
29		92	0	−92	−22.35
30	418 + 71	92	0	−581	−134.43
31		92	0	−92	−20.27
32		92	0	−92	−19.31
33		92	0	−92	−18.39
34		92	0	−92	−17.51
35		92	0	−92	−16.68
36		92	0	−92	−15.88
37		92	0	−92	−15.13
38		92	0	−92	−14.41
39		92	0	−92	−13.72
40		92	0	−92	−13.07
41		92	0	−92	−12.45
42		92	0	−92	−11.85
43		92	0	−92	−11.29
44		92	0	−92	−10.75
45	71	92	0	−163	−18.14
46		92	0	−92	−9.75

(cont.)

Table 5.10 (*Continued*)

Year	(1) Capital cost	(2) Running cost	(3) Income	(4) = (3) − (2) − (1) Net cash flow	Discount cash flow at $i = 5\%$
47		92	0	−92	−9.29
48		92	0	−92	−8.85
49		92	0	−92	−8.42
50		92	0	−92	−8.02
51		92	0	−92	−7.64
52		92	0	−92	−7.28
53		92	0	−92	−6.93
54		92	0	−92	−6.60
55		92	0	−92	−6.29
56		92	0	−92	−5.99
57		92	0	−92	−5.70
58		92	0	−92	−5.43
59		92	0	−92	−5.17
60		92	0	−92	−4.93
					NPV = −2,491.69

5.4.3 Case 3: The Toilet Flushing System Using Reclaimed Water in SD, SK, NNT and NWNT

Based on the ratio of treatment capacity shown in Table 5.7, and the capital cost, OAE and E&M capital cost of the tertiary treatment and effluent reuse treatment given in Tables 5.1 and 5.2, the same for each STW can be determined, and they are shown in Tables 5.11 and 5.12. The calculation procedures are shown next.

The capital cost for tertiary treatment is (see Table 5.1 and Eq. (5.4))

$$= \frac{(34.8101 + 13.7131) \times 12.5504}{1.0788} \tag{5.10}$$

$$= \$564.5028 \text{ million.}$$

Table 5.11 Capital investment, OAE and E&M capital cost of tertiary treatment

District of STW	Population ratio	Capital investment (HK\$ million)	OAE (HK\$ million)	E&M capital cost (HK\$ million)
SD	0.198	111.6124	6.2036	44.6450
SK	0.249	140.6348	7.8167	56.2539
NNT	0.202	114.0866	6.3411	45.6346
NWNT	0.351	198.1690	11.0146	79.2676
Total	1	564.5028	31.3760	225.8011

Table 5.12 Capital investment, OAE and E&M capital cost of effluent reuse treatment

District of STW	Population ratio	Capital investment (HK$ million)	OAE (HK$ million)	E&M cost (HK$ million)
SD	0.198	43.8252	4.9629	17.5301
SK	0.249	55.2210	6.2534	22.0884
NNT	0.202	44.7967	5.0729	17.9187
NWNT	0.351	77.8121	8.8116	31.1248
Total	1	221.6550	25.1008	88.6620

The OAE for tertiary treatment is (see Section 5.2.3 and Eq. (5.4))

$$\begin{aligned} &= \$2.5 \times 12.5504 \\ &= \$31.376 \text{ million.} \end{aligned} \tag{5.11}$$

The E&M capital cost for tertiary treatment is (see Table 5.2 and Eq. (5.4))

$$\begin{aligned} &= \frac{(13.924 + 5.4852) \times 12.5504}{1.0788} \\ &= \$225.8011 \text{ million.} \end{aligned} \tag{5.12}$$

The capital cost for chlorination/colour dyeing processes is (see Table 5.1 and Eq. (5.4))

$$\begin{aligned} &= \frac{5.8017 \times 12.5504}{0.3285} \\ &= \$221.655 \text{ million.} \end{aligned} \tag{5.13}$$

The OAE for chlorination and colour dyeing processes is (see Section 5.2.3 and Eq. (5.4))

$$\begin{aligned} &= \$2 \times 12.5504 \\ &= \$25.1008 \text{ million.} \end{aligned} \tag{5.14}$$

The E&M capital cost for chlorination and colour dyeing processes is (see Table 5.2 and Eq. (5.4))

$$\begin{aligned} &= \frac{2.3207 \times 12.5504}{0.3285} \\ &= \$88.6627 \text{ million.} \end{aligned} \tag{5.15}$$

The following is a summary of costs for Case 3.

Capital investment (see Eq. (5.1), Tables 5.11 and 5.12) = additional flushing system + tertiary sewage treatment + chlorination and colour dyeing processes = 418 + (564.5028 + 221.655) × 0.8 × 0.85	952.5873
Total running cost (see Tables 5.4, 5.5, 5.11 and 5.12) = staff cost + OAE = 1501.2835 × 1.031% + [1297.7297 × 1.031% + (31.376 + 25.1008) × 0.9]	79.6869
Renewal cost (treated effluent in SD, SK and NNT, 30 years) of flushing system (see Eq. (5.1)) = 418	418
Renewal cost (15 years) of E&M for tertiary treatment, and chlorination and colour dyeing processes (see Tables 5.11 and 5.12] = (225.8002 + 88.6627) × 0.8 × 0.85	213.8348

The NPV calculation for Case 3 is shown in Table 5.13. The NPV calculated is −HK$2740 million.

5.4.4 Summary of the Cases

A summary of the three aforementioned cases is shown in Table 5.14. The unit cost, expressed in HK$/m^3, can be derived from the NPV calculated using the following formula:

$$\text{Unit cost} = \frac{-\text{NPV} \times \{0.05(1+0.05)^{60}/[(1+0.05)^{60}-1]\}}{12.5504}.$$

The explanation of this formula can be found in Tang (2003, Chapters 1 and 2).

The cost comparison result shows that the dual water supply systems by adopting seawater for toilet flushing in these districts have the best engineering economy, then raw fresh water and then reclaimed water. The use of seawater for toilet flushing is less expensive than the other options. In addition, using seawater provides a number of benefits including reliability of supply, cutting down the demand for fresh water and reducing reliance on Dongjiang water.

Although the use of seawater for flushing has many advantages, it must be used with caution for NNT (Sheung Shui and Fanling) and NWNT (Tin Shui Wai and Yuen Long) districts. The export pipelines can help transport the saline effluent doubtlessly from the STW back to the sea without damaging the ecology, but potential danger still exists. A small portion of seawater supply may not get into the proper sewerage system after use and therefore may not be discharged into the export pipelines. This portion of saline water can be harmful to the ecology if they get into rivers or farmlands. Such a problem cannot be ignored even though only 25% of the population is assumed to be supplied with seawater for flushing in this study and the capacity of the STW in each of the districts is large enough theoretically to treat all the seawater imported.

Table 5.13 NPV calculation for Case 3

Year	(1) Capital cost	(2) Running cost	(3) Income	(4) = (3) − (2) − (1) Net cash flow	Discount cash flow at $i = 5\%$
0	953			−953	−953.00
1		80	0	−80	−76.19
2		80	0	−80	−72.56
3		80	0	−80	−69.11
4		80	0	−80	−65.82
5		80	0	−80	−62.68
6		80	0	−80	−59.70
7		80	0	−80	−56.85
8		80	0	−80	−54.15
9		80	0	−80	−51.57
10		80	0	−80	−49.11
11		80	0	−80	−46.77
12		80	0	−80	−44.55
13		80	0	−80	−42.43
14		80	0	−80	−40.41
15	214	80	0	−294	−141.42
16		80	0	−80	−36.65
17		80	0	−80	−34.90
18		80	0	−80	−33.24
19		80	0	−80	−31.66
20		80	0	−80	−30.15
21		80	0	−80	−28.72
22		80	0	−80	−27.35
23		80	0	−80	−26.05
24		80	0	−80	−24.81
25		80	0	−80	−23.62
26		80	0	−80	−22.50
27		80	0	−80	−21.43
28		80	0	−80	−20.41
29		80	0	−80	−19.44
30	418 + 214	80	0	−712	−164.74
31		80	0	−80	−17.63
32		80	0	−80	−16.79
33		80	0	−80	−15.99
34		80	0	−80	−15.23
35		80	0	−80	−14.50
36		80	0	−80	−13.81
37		80	0	−80	−13.15
38		80	0	−80	−12.53
39		80	0	−80	−11.93
40		80	0	−80	−11.36
41		80	0	−80	−10.82
42		80	0	−80	−10.31
43		80	0	−80	−9.82
44		80	0	−80	−9.35
45	214	80	0	−294	−32.72
46		80	0	−80	−8.48

(cont.)

Table 5.13 (*Continued*)

Year	(1) Capital cost	(2) Running cost	(3) Income	(4) = (3) − (2) − (1) Net cash flow	Discount cash flow at $i = 5\%$
47		80	0	−80	−8.08
48		80	0	−80	−7.69
49		80	0	−80	−7.33
50		80	0	−80	−6.98
51		80	0	−80	−6.64
52		80	0	−80	−6.33
53		80	0	−80	−6.03
54		80	0	−80	−5.74
55		80	0	−80	−5.47
56		80	0	−80	−5.21
57		80	0	−80	−4.96
58		80	0	−80	−4.72
59		80	0	−80	−4.50
60		80	0	−80	−4.28
					NPV = −2,740.33

As to the use of raw water or reclaimed water, although using the latter is more expensive than the former at present, it is predicted that the situation will reverse after 2010 if the raw water price increases in the same way as it did in the past 10 years. From the results of Cases 2 and 3, the NPVs are −HK$2492 million and −HK$2740 million, respectively. The current unit price of raw water purchased from Dongjiang is HK$3.085/m^3. When this unit price increases and reaches a certain level, the raw water alternative would no longer be attractive. Repeated trial calculations have been carried out by adjusting the unit price of raw water, and it is obtained that when the unit price increases to HK$4.25 (year 2000 money value based) the NPV of Case 2 would become −HK$2740 million, which is equal to that of Case 3.

According to the WSD's annual report (WSD 2003d), the real unit price of raw water from Dongjinag has gradually increased in the last decade as shown

Table 5.14 Summary of Cases 1–3

	Description	NPV (HK$ million)	Unit cost (HK$ per m^3)*
Case 1	Seawater in SD, SK, NNT and NWNT	−1,398	HK$5.88
Case 2	Raw water in SD, SK, NNT and NWNT	−2,492	HK$10.49
Case 3	Reclaimed water in SD, SK, NNT and NWNT	−2,740	HK$11.53

Table 5.15 Real unit cost of raw water from Dongjiang

Year	CPI	Unit cost of raw water (HK$)	Real unit cost of raw water at 2000 basis (HK$)
1993	78.3	1.772	2.2631
1994	85.2	1.94	2.2770
1995	92.9	2.16	2.3251
1996	98.8	2.405	2.4342
1997	104.5	2.613	2.5005
1998	107.5	2.839	2.6409
1999	103.2	3.085	2.9893
2000	99.4	3.085	3.1036
2001	97.8	3.085	3.1544
2002	94.8	3.085	3.2542

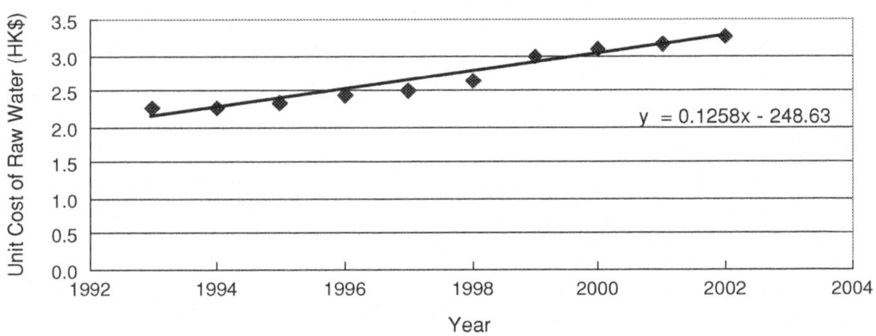

Figure 5.3 Trendline of real unit cost (year 2000 based) of raw water from Dongjiang.

in Table 5.15. A trendline is determined by linear regression based on the data in Table 5.15 and is shown in Figure 5.3. The regression line implies that when 2010 is reached, i.e. when the unit price of raw water becomes HK$4.25, using raw water and reclaimed water would have the same NPV.

5.5 COMPARISON OF THREE MORE CASES

There are three more cases that are worthy of comparison. They are

Case 4: The toilet flushing system using seawater in SD and SK and raw water in NNT and NWNT.

Case 5: The toilet flushing system using seawater in SD and SK and reclaimed water (treated effluent) in NNT and NWNT.

Case 6: The toilet flushing system using seawater in SD and SK, raw water in NNT and reclaimed water (treated effluent) in NWNT.

5.5.1 Summary of Costs for Case 4

The following is a summary of costs for Case 4.

Capital investment (see Eq. (5.1) and Table 5.12) = additional flushing system + chlorination/colour dyeing processes = 418 + (44.7967 + 77.8121) × 0.8	516.087
Total running cost (see Tables 5.4, 5.5, 5.9 and 5.12) = staff cost + OAE + water purchase cost = 1501.2835 × 1.031% + [1297.7297 × 1.031% + (5.0729 + 8.8116) × 0.9] + (8.2541 + 14.3374)	63.9454
Renewal cost (seawater, 20 years) of flushing system (see Table 5.6) = 82.764 + 104.082	186.846
Renewal cost (raw water, 30 years) of flushing system (see Table 5.6) = 84.436 + 146.718	231.154
Renewal cost (15 years) of E&M capital costs for chlorination and colour dyeing processes (see Table 5.12) = (17.9187 + 31.1248) × 0.8	39.2348

The NPV calculated for Case 4 is −HK$1910 million. The calculation is not shown and is left for readers to verify.

5.5.2 Summary of Costs for Case 5

The following is a summary of costs for Case 5.

Capital investment (see Eq. (5.1) and Tables 5.11 and 5.12) = flushing system + tertiary treatment + effluent reuse treatment = 418 + (114.0866 + 198.169) × 0.8 × 0.85 + (44.7967 + 77.8121) × 0.8 × 0.85	713.7078
Total running cost (see Tables 5.4, 5.5, 5.11 and 5.12) = staff cost + OAE = 1501.2835 × 1.031% + [1297.7297 × 1.031% + (6.3411 + 11.0146) × 0.9 + (5.0729 + 8.8116) × 0.9]	56.974
Renewal cost (seawater, 20 years) of flushing system (see Table 5.6) = 82.764 + 104.082	186.846
Renewal cost (treated effluent, 30 years) of flushing system (see Table 5.6) = 84.436 + 146.718	231.154
Renewal cost (15 years) of E&M capital costs for tertiary treatment, and chlorination and colour dyeing processes (see Tables 5.11 and 5.12) = (45.6346 + 79.2676) × 0.8 × 0.85 + (17.9187 + 31.1248) × 0.8 × 0.85	118.2831

The NPV calculated for Case 5 is −HK$2041 million. The calculation is again not shown and is left to the readers to verify.

5.5.3 Summary of Costs for Case 6

The following is a summary of costs for Case 6.

Capital investment (see Eq. (5.1) and Tables 5.11 and 5.12) = additional flushing system + chlorination/colour dyeing processes $= 418 + 198.169 \times 0.8 \times 0.85 + (44.7967 + 77.8121) \times 0.8 \times 0.85$	636.1289
Total running cost (see Tables 5.4, 5.5, 5.9, 5.11 and 5.12) = staff cost + OAE + water purchase cost $= 1501.2835 \times 1.031\% + [1297.7297 \times 1.031\% + (11.0146) \times 0.9 + (5.0729 + 8.8116) \times 0.9] + 8.2541$	59.5211
Renewal cost (seawater, 20 years) of flushing system (see Table 5.6) $= 82.764 + 104.082$	186.846
Renewal cost (raw water, 30 years) of flushing system (see Table 5.6) $= 84.436 + 146.718$	231.154
Renewal cost (15 years) of E&M capital cost for chlorination and colour dyeing processes (see Tables 5.11 and 5.12) $= (79.2676) \times 0.8 \times 0.85 + (17.9187 + 31.1248) \times 0.8 \times 0.85$	87.2515

The NPV calculated for Case 6 is −HK$1993 million.

5.5.4 Summary of Cases 4–6

A summary of the three aforementioned cases is shown in Table 5.16. The unit cost, expressed in HK$ per m^3, can be derived from the NPV calculated using the same formula as given in Section 5.4.4.

As discussed in Section 5.4.4, when the unit price of raw water becomes HK$4.25, using raw water and reclaimed water would have the same NPV. Similar

Table 5.16 Summary of Cases 4–6

	Description	NPV (HK$ million)	Unit cost (HK$ per m^3)*
Case 4	Seawater in SD and SK, and raw water in NNT and NWNT	−1910	HK$8.04
Case 5	Seawater in SD and SK, and reclaimed water in NNT and NWNT	−2041	HK$8.59
Case 6	Seawater in SD and SK, raw water in NNT and reclaimed water in NWNT	−1993	HK$8.39

result can be obtained from comparing Cases 4 and 5, the NPVs of which are −HK$1910 million and −HK$2041 million, respectively. The same analysis (see Section 5.4.4) is carried out to obtain the break-even point. By increasing the unit price of raw fresh water to HK$4.25 (year 2000 money value based) for Case 4, the NPV would become −HK$2041 million. This implies again that when 2010 is reached, using raw water and reclaimed water would have the same NPV. Of course, this is true only if the raw water price increases in the same way as it did in the past 10 years. After that time, using reclaimed water, besides being cheaper than raw fresh water, can also help reduce the reliance on Dongjiang water.

5.6 REFERENCES

Advisory Committee on the Quality of Water Supply (ACQWS). 2003, ACQWS Paper 14 – Treated Effluent Reuse at Ngong Ping. Water Supplies Department, Hong Kong.

Census and Statistical Department (CSD). 2003, Consumer Price Indices. The Gorvernment of Hong Kong SAR, Hong Kong.

Tang, S. L. 2003, Economic Feasibility of Projects: Managerial and Engineering Practice. 3rd edition, Chinese University Press, Hong Kong.

Tang, S. L., Tsang, W. L. and Leung, H. Y. 2003, A Comparative Study of the Economy of 'Single Water Supply System' and 'Dual Water Supply Systems'. Research Report, The Hong Kong Polytechnic University, Hong Kong. Available in the Hong Kong Polytechnic University's library.

Tang, S. L., Yue, Derek, P. T., and Li, X. Z. 2006, Comparison of Engineering Costs of Seawater, Reclaimed Water and Fresh Water for Toilet Flushing in Hong Kong. Water and Environment Journal, Journal of CIWEM, (Vol. 20, No. 4.)

Water Supplies Department (WSD). 2003a, Homepage of WSD, The Government of Hong Kong SAR (www.info.gov.hk/wsd/indexe.htm).

Water Supplies Department (WSD). 2003b, My water – Homepage of WSD, the Government of Hong Kong SAR (http://www.info.gov.hk/wsd/en/html/water/hkwchn.htm)

Water Supplies Department (WSD). 2003c, Operating Accounts for the Year Ended 31 March 2003. The Government of Hong Kong SAR.

Water Supplies Department (WSD). 2003d, Water Supplies Department Annual Report 2002-2003. The Government of Hong Kong SAR.

6

Possibility of using urban runoff as toilet flushing water

6.1 ADVANTAGES OF USING URBAN RUNOFF FOR TOILET FLUSHING

Urban runoff is the surface water running on the urban area resulting from the rainfall precipitation. Although a portion of precipitation is absorbed into the ground or evaporated during the transportation, the remaining portion is large and can be collected by the drainage system in an urban area. The amount of urban runoff is largely a factor of duration and intensity of rainfall and the amount of paved area in a city.

As the city grows and more land area is developed under the urbanisation, it modifies the pervious surface in natural terrain to impervious like roads, highways, bridges, car parks, gardens, buildings, etc. It changes the rainfall-runoff process in a variety of ways. Infiltration is reduced due to the addition of imperious surfaces, resulting in increasing quantities of runoff. Tree removal, surface levelling, soil flipping and surface compaction area also likely to boost the quantity of runoff. In addition, the rate of storm-water runoff is intensified due to the extensive collection networks of pipes and channels provided in the urban environment. The long surface travel time from previous undeveloped land is thus shortened, and gutters and pipes quickly convey stormwater to receiving streams, rivers or sea (Akan and Houghtalen 2003).

The urban drainage system concentrates runoff into smooth and straight underground conduits. The runoff gathers speed and erosional power when travelling. When this runoff leaves the storm drains and empties into a stream, its excessive volume and power may blast out streambanks, damaging streamside vegetation and wiping out aquatic habitat (US EPA 2003). The increased storm flows carry a large amount of sediment and pollutant loads from the urban areas. They often carry higher water temperatures from streets, rooftops and parking lots, which are harmful to the health and reproduction of aquatic life. The loss of infiltration from urbanisation may cause profound groundwater changes.

Clogged storm drains can lead to area flooding when it rains, creating traffic problems and unsanitary conditions. Hence, the citizens generally opine that urban runoff is a nuisance or may pose a risk to their property. The design approach of storm-water drainage system is therefore to divert surface runoff from inland or local streams as far as possible in order to protect the public from the dangers and inconvenience of flooding.

In most urban water supply systems, the raw water is obtained from the clear water courses at a great distance away. Cities require constructing a huge conveyance system to deliver large volumes of high-quality water into urban areas for daily usage. As the cities have the ability to divert the urban runoff via the drainage system in order to prevent flooding, they should therefore have the ability to make use of the existing system to collect and divert the relatively good quality runoff to a place for storing and using later. In the recent decades, town planners and engineers have begun to explore alternatives to the runoff disposal, such as the use of stormwater for toilet flushing (Mitchell *et al.* 1999). This approach offers many benefits including the following:

(1) The urban runoff is a costless resource.
(2) The collected urban runoff can be the water source providing at or near the area where the flushing water is needed.
(3) The urban runoff toilet flushing system can co-exist with the potable water supply system, which provides a good supplement to the valuable potable resource.
(4) The urban runoff can be extensively collected from the urban area via the existing storm-water collection system in urban area with few negative environmental impacts compared to other wastewater reuse technologies.
(5) The urban runoff collection system can reduce storm drainage load and flooding in city streets.
(6) Cities do not have to rely on importing large volume of high-quality water from the surrounding water sources such as rivers and groundwater.
(7) The relatively good quality runoff would require less treatment than the potable water, especially for the toilet flushing purpose that do not require a stringent water quality as the water is not likely to come into human contact.
(8) The use of urban runoff can reduce the volume of water and peak flows in watercourses. By levelling out and reduce excessive peak flows, the flushing effects and damages to the river and streams during flooding can be minimised.

6.2 URBAN RUNOFF COLLECTION SYSTEM

For a typical storm-water drainage system, it should be designed to collect and convey surface runoff generated within cities during and after rainfall events, for safe discharge into a receiving watercourse or the sea. It is to protect the public from the dangers and inconvenience of flooding. The system typically comprises surface drainage, road kerb gullies, drain pipes, open or underground channels and culverts. The magnitude of peak flows that have to be accommodated depends primarily on the intensity of rainfall and the size, topography, soil type, configuration and land use of the catchment.

The components of the conveyance system in the urban runoff collection system are similar to typical drainage system including pipes, channels and culverts. However, it requires transferring the collected runoff to storage facilities. The runoff then has to be treated to comply with the defined water quality standard for toilet flushing before delivering to the community through a separate distribution system.

6.2.1 Conveyance System

When designing a conveyance system, consideration should be given to dirt and debris from the catchment collected from the first part of rainfall. Relatively clean water will only be available sometime later in the storm. The initially collected runoff shall therefore be diverted from the storage facilities. The common method is providing sediment trap in the road kerb gullies or in the intermediate manholes along the drain-pipe alignment in order to prevent the entry of dirt and debris from the catchment surface. The pipes used for conveying rainwater should be made of concrete, plastic, PVC or other inert substance wherever possible as the acidic rainwater may cause corrosion to the metal pipes.

6.2.2 Storage Facilities

The storage facilities for collecting rainwater can be located in or away from urban areas. Various types can be designed and constructed to accommodate the town planning and environmental consideration. Structural storage tanks and retaining reservoirs are the common types adopted in cities. They are usually constructed with reinforced concrete due to its inert material properties. In addition, the structures are easy to clean and can feasibly provide openings to fit for the incoming and discharging pipes.

Intermediate urban runoff control facilities can also serve the storage purpose such as flood detention basins and wet ponds. They not only capture the runoff but also improve the water quality by removing suspended matters and associated constituents such as bacteria, nutrients and metals by means of sedimentation.

6.2.3 Treatment Processes for Urban Runoff

Water treatment is an inevitable process in the urban runoff reuse system. To ensure that the reclaimed effluent is suitable for toilet flushing, the impurities or pollutants in the runoff such as oil, grease, large solid, organic and inorganic compounds, bacteria and pathogens, etc. shall be removed (Antich *et al.* 2002).

The preliminary treatment of runoff is necessary for the proper functioning of downstream processes in the treatment plant. It can consist of flow-equalisation basin to stabilise the runoff flows, bar screens and flotation units to remove large debris, oil and grease. These units were deemed best from the viewpoint of space constraints, operation and maintenance, and residuals management. Then the runoff goes to undertake secondary treatment processes.

Secondary treatment processes may use fine screening, sand filtration, microfiltration, lime softening, reverse osmosis, granulated activation carbon and ion exchange filters. Each of these systems is capable for removing the contaminations. Available space of treatment plant, residual management and familiarity of operation and maintenance shall be evaluated when selecting the system.

Several disinfection processes are applicable including the use of ozone, UV radiation and sodium hypochlorite in the treatment plant. Selection criteria include the plant and community safety, regulatory acceptance, pilot experience, operation and maintenance consideration, organic removal rate, cost and environmental acceptance. Although all the suggested processes can satisfy the concerns, the UV process is commonly selected because of its small physical site layout requirements, minimal chemical handling, reduced environmental impact and relatively low costs.

6.2.4 Distribution System

A separate distribution system shall be designed and constructed to deliver the treated runoff for toilet flushing. Colourised water pipes with permanent identity marking should be used to avoid confusing or misconnection with the potable water distribution system. The peak period for using the treated runoff is during the night-time in weekday, and the whole day in weekend or public holiday. However, a relatively small peaking factor is applied for the system design due to the low variability of flow demand.

6.3 WATER QUALITY OF URBAN RUNOFF

In the past, it was believed that rainwater was pure and could be consumed without pre-treatment. This may be true in some areas that are relatively unpolluted. However, with the rapid economic growth and urbanisation, rainwater collected in many locations contains impurities. Particularly during the last three decades, 'acid rain' occurs as a result of sulphur dioxide, nitrous oxide and other industrial emissions. These have affected the quality of rainwater (United Nations Environment Programme 2002).

Pollutants accumulated on the land surface of an urban area during dry periods. Under the rainstorm, these pollutants are washed off the land surface and transported with the runoff. The University of New South Wales (UNSW) Research Report No. 204 entitled *Storm Water Quality from Road Surfaces – Monitoring of Hume Highway at South Strathfield* provides some information on the pollutant concentration from roads. Key findings of the report are as follows:

(1) It has been found that dry-weather build-up of contaminants reaches equilibrium after 10 days. After 10 days, contaminant deposition rate will then be similar to removal rate caused by air turbulence. A probable reason for this is that the contaminant can stick to the road but once there is a layer of contaminant they cannot stick to each other any longer.
(2) It has been found that the contaminant equilibrium can be maintained until a cleansing event occurs. Cleansing events are defined as wind events exceeding 21 km/hour or storm events with rainfall exceeding 7 mm/hour.
(3) Contaminant concentrations in runoff are negligible after 20–30 minutes.
(4) Lead concentration is low and in the order of micrograms (10^{-6}) per litre. Although the instant peak in the initial period of runoff collection is 400 μg/L or 0.4 mg/L, the concentration in terms of a rainfall event is practically nil.

Urban storm-water quality is influenced by many factors, including population density, land use, relative proportion of impervious area, waste-disposal and sanitation practices, soil type, climate and local construction activity. The interaction of these factors contributes to the variable quality of stormwater.

Although some types and quantities of pollutants vary among cities, all have common pollutants entering the drainage collection systems via cars, roads and construction. Table 6.1 shows the common pollutants.

Urbanisation increases the variety and amount of pollutants flowing into the drainage system. The deteriorated quality of urban runoff not only affects the treatment cost but may also impose health risk to public. These pollutants will be further discussed in the next section.

6.4 POLLUTANTS IN URBAN RUNOFF

6.4.1 Suspended Solids

Suspended solids in urban runoff most likely originate from erosion at construction sites or the wash-off of particles that build up on impervious surfaces during dry periods. The solids deposited in receiving waters may limit light penetration (and thus photosynthesis) or upset aquatic reproductive processes. In addition, other pollutants often become attach to the suspended solids and are carried in urban runoff (Akan and Houghtalen 2003). It is therefore generally assumed that there is a relatively constant ratio between the suspended solid load and other pollutants

Table 6.1 Comparison of contaminant profiles for the urban surface runoff and raw domestic sewage, based on surveys throughout USA (by the University of Michigan)

Constituent	Urban surface runoff	Raw domestic sewage
Suspended solids	250–300	150–250
BOD*	10–250	300–350
Nutrients		
Total nitrogen	0.5–5	25–85
Total phosphorus	0.5–5	2–15
Coliform bacteria (MPN†/100 mL)	10^4–10^6	10^4 or greater
Chorides	20–100	15–75
Oil and grease	Yes	Yes
Heavy metals	(10–100) times sewage concentration	Traces
Pesticides	Yes	Seldom
Other toxins	Potential exists	Seldom

All concentrations are expressed in mg/L unless stated otherwise.
*Biochemical oxygen demand.
†Most probable number.

in the runoff. Suspended solid concentration is often used as a primary indicator of stormwater pollution.

6.4.2 Nutrients

Nutrients, which are primarily phosphorus and nitrogen, come from lawn fertilisers. It promotes the growth of algae in receiving waters. At elevated levels, algae upset the normal ecosystem balance, and the death and decay of algae depletes the water's oxygen supply.

6.4.3 Faecal Coliform Bacteria

Faecal coliform bacteria may be present during and immediately following storm events. The coliform bacteria by themselves are not pathogenic. That means they do not cause illness or disease. If coliform bacteria are present in amounts greater than 2000 colonies per millilitre of water, however, it is likely that other pathogenic organisms will also be present such as other bacteria, viruses and parasites that do cause disease and illness (Hall 2005).

6.4.4 Heavy Metals

Heavy metals (e.g. Cu, Cr, Zn, Pb and Cd) come from automobiles gutters, and building materials are common pollutants in wet-weather flows and urban waterways (Pitt *et al.* 1995). Unlike organics, metals are not degraded in the environment.

They can exert short-term toxicity and, because of accumulation in sediments, long-term toxicity as well. Metals have been shown to associate with particles in the stormwater.

6.4.5 Other Pollutants

Other pollutants include gasoline and oil from motor vehicles, pesticides, insecticides and herbicides from lawn and shrub care products, litter and solid waste from people and faecal droppings from animals. These shall be taken into consideration during the collection and carefully removed from the urban runoff in the treatment system for toilet flushing purpose.

6.5 PRECAUTIONARY MEASURES FOR PROTECTING THE RUNOFF QUALITY

6.5.1 Government of a City

To safeguard the quantity of runoff collected, the government of a city should look for priority pollutant reduction opportunities in the planning stage. With reference to the experience from Singapore, the following precautionary measures (Lee and Haja 1996) can be taken in town planning:

(1) The pollutive industries can be relocated outside the new town.
(2) All the drains and gutters around the residential buildings are best covered up.
(3) New development and re-developments in catchment should be carefully planned at the planning stage to ensure they are non-polluting and compatible with developments allowed in water catchment.
(4) Suitable design and operation of the stormwater collection system are required to ensure that only the cleaner part of stormwater is abstracted and pumped to collection ponds.
(5) Drains and sand traps are more regularly cleansed to avoid the accumulation of debris.
(6) The use of pesticide/herbicide in the parks of catchment should be minimised, biodegradable and applied cautiously.
(7) Provision of design such that runoff from markets and food stalls is completely intercepted to foul sewer.

6.5.2 Developers

The development in the catchment area should be carefully planned and designed by the developers and the engineers. They should propose and implement pollution prevention strategies and examine the source control measures in order to effectively avoid or minimise the potential pollution to the urban runoff. Low-impact development design concept should also be adopted including the measures that conserve natural areas, maintain local ecological environment and retrofit activities to clean up existing degraded water bodies. The developers can require the *contractor* by means of *contract* provision to reduce the polluted runoff, under the construction stage, to flow into the collection system (US EPA 2003).

6.5.3 House Owners

To decrease polluted runoff from paved surfaces, households can develop alternatives to areas traditionally covered by impervious surfaces. Porous pavement materials are available for driveways and sidewalks, and native vegetation and mulch can replace high maintenance grass lawns. Home owners can use fertilisers sparingly and sweep driveways, sidewalks and roads instead of using a hose. Instead of disposing of yard waste, they can use the materials to start a compost pile.

In addition, households can prevent polluted runoff by properly using, storing and disposing of chemicals properly. Households served by septic tank systems should have them professionally inspected and pumped every 3–5 years. They should also practice water conservation measures to extend the life of these systems.

6.5.4 Vehicle Drivers

Drivers should check their cars for leaks and recycle their motor oil and antifreeze when these fluids are changed. Drivers can also avoid impacts from car wash runoff (e.g. detergents, grime, etc.) by using car wash facilities that do not generate runoff.

6.5.5 Construction Sites

Construction sites should control sediment-laden runoff from disturbed areas during the construction phase of development. All stormwater runoff from hard surfaces should be treated through infiltration, sedimentation, chemical or biological treatment before leaving the site. Minimal sediment generation and erosion control measures during construction should also be implemented. Stormwater directly discharged to waterways without treatment is prohibited.

6.6 CASE STUDIES ON URBAN RUNOFF COLLECTION

6.6.1 Case Study in Singapore

Singapore is a small country with a total land area of approximately 600 km^2. About 43% of the land area has been developed as residential, commercial and industrial areas and 15% is agricultural land. The remaining areas are either natural reserves or other non-built-up areas. The average annual rainfall is 2400 mm, which is also the main source of water supply. As water is a scarce and limited resource in Singapore, almost half of the total land area has been developed as water catchment areas for the collection of rainwater into 14 impounding reservoirs (Chui 1991).

Figure 6.1 Location map of Lower Seletar and Bedok Reservoirs/stormwater.

To meet the increasing demand for water, the Public Utilities Board of Singapore started urban stormwater pond collection systems in suitable catchments in 1986. It is to exploit the harnessing of storm runoff from new townships at the north-eastern part (Seletar) and eastern part (Bedok) of Singapore to augment its limited water supplies (see Figure 6.1). The scheme has been in use over 20 years.

The stormwater pond collection systems was implemented to harness the surface runoff from built-up residential areas of Ang Mo Kio, Bedok, Tampines and Yishun New Towns as a source of raw water. The surface runoff collected is then channelled into the Lower Seletar and Bedok Reservoirs, as shown in Figure 6.1, via a complex collection and conveyance system comprising nine collection ponds, nine pumping stations, connecting pipelines and a gravity flow diversion channel (Lim and Lim 1999).

6.6.1.1 Design of Collection System

Rainfall in Singapore is characterised by relatively high-intensity and short-duration storm events of small aerial distribution and moving storms. The implemented urban stormwater collection system is therefore designed to respond quickly to collect a large volume of water flow in a relatively short time.

The chosen configuration consists of a diversion structure that channels the flash flow into a holding pond. The structure consists of a grating section installed in the channel floor of an existing stormwater drain. A dry-weather flow channel runs past the grating section and discharges to the sea. During storm events, when the dry-weather flow-channel capacity is exceeded, the rising water level will cause

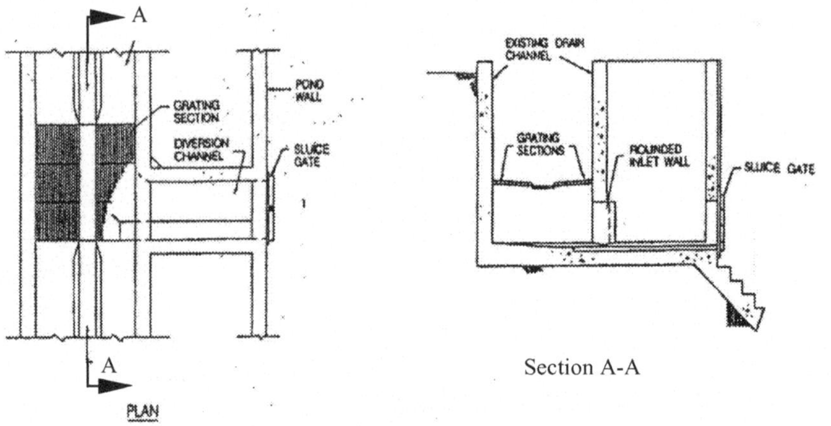

Figure 6.2 Typical section of open drop inlet structure. *Source*: Lee and Haja 1996.

the runoff to spill through the grating section to a sump below. This structure prevents low-quality dry-weather flow and the first flush of a storm flow, which usually contains accumulated pollutants, from entering the drop inlet. Diverted flow leaves the sump through a sloping diversion channel to a detention pond. At the entrance to the pond, a sluice gate is provided to enable the pond to be isolated for maintenance works (see Figure 6.2).

Another structure, as shown in Figure 6.3, for diverting stormwater flows into a detention pond or a reservoir is more dynamic. A barrage gate is installed in the

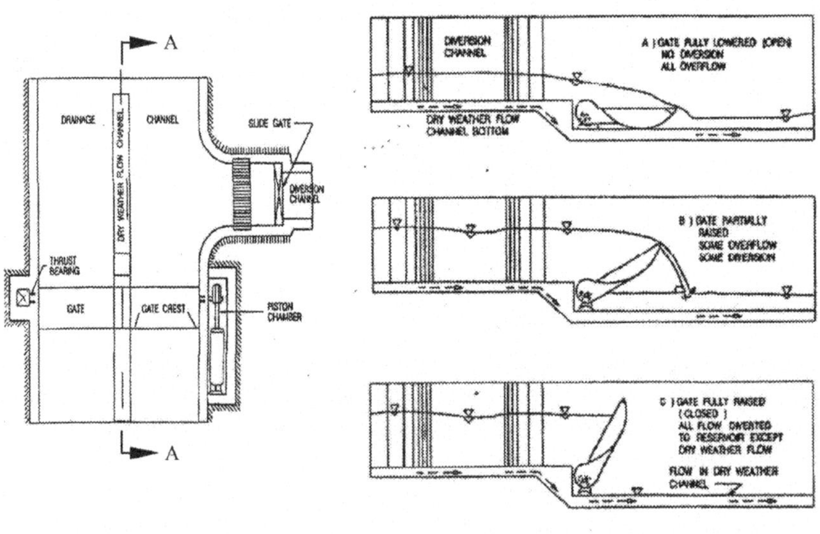

Section A-A

Figure 6.3 Typical plan and section barrage gate lateral diversion system. *Source*: Lee and Haja 1996.

Table 6.2 Typical quality of collected stormwater (1988–1994)

Parameter	Range	Average
Conductivity (umhos/cm)	40–20,900	275
Total organic carbon (mg/L)	1.4–35.8	7.8
Ammoniacal nitrogen (mg/L)	<0.02–20	0.33
Nitrate nitrogen (mg/L)	<0.1–3.62	0.37

main drainage channel just downstream of the diversion point. The gate extends the full width of the drain and is rotated about a torque tube anchored to the drain bottom. The position of the gate is controlled by a hydraulic cylinder located in a dry well adjacent to the drain, providing a 60° range of gate travel measured from the horizontal. In the open and lowered position, the gate will be in a horizontal plane. The drainage channel immediately downstream of the barrage gate is provided as a stilling basin to dissipate the energy of large spills over the crest of the gate. A dry-weather flow channel is incorporated beneath the barrage gate and through the stilling basin sill.

6.6.1.2 *Water Quality of Collected Runoff*

The quality of stormwater collected in the detention ponds is shown in Table 6.2 as to the conductivity, total organic carbon, ammoniacal nitrogen and nitrate nitrogen. The conductivity in the stormwater is largely due to some tidal influence as the catchments are generally low-lying. TOC consists of mainly particulate organic carbon, which is readily removed by storage in the reservoirs. The ammonical nitrogen and nitrate nitrogen are, on average, lower than those of raw water abstracted from some river resources (Lee and Haja 1996).

Bedok Reservoir was developed from a huge quarry. It does not have its own direct catchment and serves as a storage reservoir receiving pumped flows from Lower Seletar Reservoir and nine urban stormwater collection stations. Table 6.3 shows the typical quality of the Bedok Reservoir and Lower Seletar Reservoir water in 1996. The quality of raw water collected from the urbanised catchments and stored at the reservoirs has generally been good as compared to raw water from largely forested catchments (Lim and Lim 1999).

6.6.1.3 *Water Treatment Process*

Bedok Waterworks was designed with a high degree of automation and mechanisation. Pro-ozonation is used to improve the coagulation-flocculation process for effective removal of organic substances and better clarification. Dissolved iron and manganese are oxidised into suspended particles and removal by settling out in the clarifiers. The TOC level in the water after clarification is below 2 mg/L

Table 6.3 Typical water quality at Bedok Reservoir and Lower Seletar reservoir

Parameter	Bedok	Lower Seletar
Physical examination		
Colour (Hazen unit)	23	80
Turbidity (NTU)	3.9	28.6
Conductivity (umhos/cm)	268	339
pH value	8	7.8
Chemical examination (mg/L)		
Chloride (as Cl)	37	51
Sulphate (as SO_4)	19	23
Total alkalinity (as $CaCO_3$)	48	57
Total hardness (as $CaCO_3$)	64	74
Ammoniacal nitrogen (as N)	0.04	<0.02
Phosphate (as PO_4)	<0.05	0.1
Total organic carbon (as C)	3.1	3.5
Iron (as Fe)	0.04	1.2
Manganese (as Mn)	0.005	0.006
Total coliform	119	62

(as carbon). The treated water has a low chlorine demand. The sludge from the sedimentation process is treated by gravity thickening, conditioning with lime and dewatered by mechanical filter presses, and the sludge cakes is then disposed of landfill in Government disposal grounds. The typical treated water quality is shown in Table 6.4. The treated water fully meets the latest WHO Guidelines for Drinking Water Quality (2006). Bedok Waterworks has successfully treated raw water trapped from urbanisation catchments and has been supplying the treated water to the eastern part of Singapore (Bedok) for 20 years since its commissioning in 1986.

So, the urban runoff collected in the Bedok case is suitable for not only toilet flushing but also for potable use. If the purpose of collecting the runoff is for toilet flushing alone, the treatment process can be simpler so as to achieve a balance of the cost and the intended use.

6.6.2 Case Study in USA (Santa Monica)

6.6.2.1 Project Background

Santa Monica is about 20.5 km^2 in size with a residential population of about 90,000. The City is surrounded by the Pacific Ocean (Santa Monica Bay) on the west, Santa Monica Mountains to the north and cities of Los Angeles and Venice to the east and south. Attractive beaches and the Santa Monica Pier, pleasant year-round climate and proximity to attractions in Southern California make Santa Monica a popular city (Shapiro 2003).

Table 6.4 Typical quality of Bedok-treated water (typical values)

Parameter	Treated water	WHO guideline values (2006)
Physical characteristic		
Colour (Hazen unit)	<5	15
Turbidity (NTU)	<1	5
Conductivity (umhos/cm)	250–550	—
pH value	7–8.5	—
Taste	Unobjectionable	—
Odour	Unobjectionable	—
Chemical characteristic (mg/L)		
Nitrate nitrogen (as N)	0.1–1.5	<50
Fluoride (as F)	0.5–0.7	1.5
Total dissolved solid	200–350	1000
Total alkalinity (as $CaCO_3$)	20–50	—
Total hardness (as $CaCO_3$)	50–100	—
Chloride (as Cl)	35–100	250
Sulphate (as SO_4)	30–60	250
Phosphate (as PO_4)	<0.1	—
Silica (as SiO_2)	1–10	—
Iron (as Fe)	<0.04	0.3
Manganese (as Mn)	<0.05	0.4
Copper (as Cu)	<0.05	2
Aluminium (as Al)	<0.1	0.2
Biological characteristic		
Coliform organisms/100 mL (35°C, 24 hours)	Not detected	Not detected

According to a fact sheet published by Environmental Programs Division of Santa Monica (2006), runoff is the single greatest source of water pollution in the Santa Monica Bay. The polluted runoff increases health risks to swimmers near the storm drain outlets and toxicity to aquatic life. These impacts translate into losses to the tourism economy, losses of recreational resources, dramatic cost increases for cleaning up contaminated sediments, degraded water quality and impaired function and vitality of natural resources.

The City of Santa Monica adopted a strategy to solve both problems: harvest stormwater, treat it and infiltrate back into the ground, and keep a pollution source out of the Bay. The City's comprehensive watershed-urban runoff management approach includes (1) an ordinance to require the harvesting of stormwater runoff from new development, (2) a philosophy of treating all dry weather and some wet-weather urban runoff before discharge or reuse and (3) a first of its kind innovative recycling facility for dry-weather runoff.

Under the approach, the Santa Monica Urban Runoff Recycling Facility (SMURRF), as shown in Figure 6.4, began operation in December 2000. It is the first full-scale, dry-weather runoff recycling facility in the USA. The facility treats low-flow dry-weather runoff from Pico-Kenter and Pier storm drains

Figure 6.4 Santa Monica urban runoff recycling facility.
Source: http://santa-monica.org/epwm/smurrf/smurrf.html. Accessed on 2006.

on a year-round basis and reuses the water for irrigation and toilet flushing. City irrigation sites include two parks, a cemetery, a middle school, several greenbelt roadway medians, the civic centre area and major office building complex (Antich *et al.* 2002).

The SMURRF harvests on and treats both wet-weather (rainwater) or dry-weather (non-sewage wastewater; see Section 6.6.2.2) runoff since the commencement of the facility. About 93% on an annual basis of dry-weather urban runoff is treated which is collected from two main storm drain lines in the city.

6.6.2.2 Water Quality of Collected Runoff

Runoff captured by the SMURRF originates from a 153-km^2 drainage area in the cities of Santa Monica and Los Angeles. Sources of dry-weather runoff arise from the inefficient use of potable water by people: over-irrigation, broken irrigation systems, washing of paved surfaces and business equipment, car washing on hard surfaces, pool draining, leaking water pipes and hydrants, and illegal dumping. The average daily flow is estimated to be 1100 m^3 per day, which represents slightly more than 2% of Santa Monica's overall water demand of 49,000 m^3 per day. The facility has a capacity of 1900 m^3 per day. A variety of pernicious contaminants are found in urban runoff. The presence and concentration of these contaminants appear to vary significantly over time. They include suspended solids, oil and grease, trash and other debris, pathogens, organics and heavy metal (lead, copper, zinc and chromium) (Shapiro 2003).

6.6.2.3 Water Treatment Process in SMURRF

As shown in Figure 6.5, runoff is first screened in a continuous deflective separation (CDS) unit. Coarse screening removes large floating debris and trash that typically flow down to open drains on streets. The screened flow is then pumped about half a mile to the recycling facility, which can process up to 1900 m^3 per day of runoff.

The first process unit in the plant is the rotating drum screen. It is used to remove the fine floating particles. The effluent will then flows to the cyclone-type grit chamber. This unit removes the grit and sand. Screening and de-gritting systems remove inorganic settleable material that may damage the downstream treatment processes and reduce the treatment efficiency.

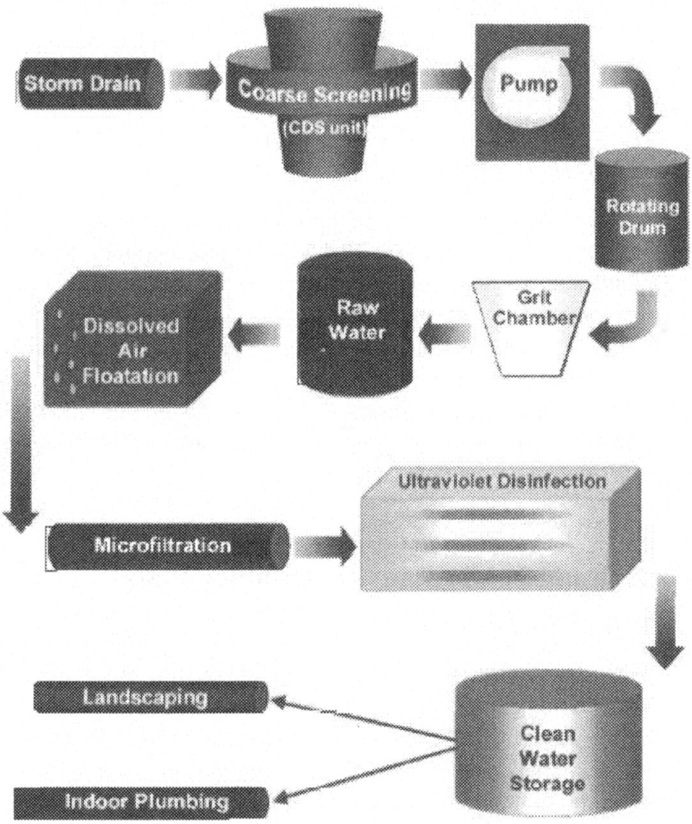

Figure 6.5 Process flow of SMURRF. *Source*: CH2M Hill Ltd 2006.

From the grit unit, the primarily treated water is stored in the raw water storage tank. The raw water storage tank dampens the fluctuations in the influent flows, thereby allowing downstream filtration and disinfection processes to operate at a steady rate, and more efficiently.

From the raw water storage tank, the water is pumped to a dissolved air flotation (DAF) unit. Compressed air is injected into the water at the unit inlet. As the water reaches the open tank surface, it reaches atmospheric pressure whereby fine air bubbles are released in the water. The air bubbles rise to the top and carry with it the oil and grease. The oil and grease blanket formed on the DAF unit open surface is then skimmed off at the top.

From the DAF unit, the water flows to the microfiltration treatment units. Here, the vacuum applied on membranes forces the water to go through the membrane thereby filtering out the turbidity. The membranes have to be periodically cleaned of the pollutant build-up.

From the microfiltration unit, the water flows to the disinfection process. Ultraviolet (UV) radiation disinfects the water. As the water passes by the UV lamps in series, UV light kills bacteria and viruses. The UV bulbs have to be periodically cleaned or replaced to prevent efficiency loss. Finally, treated water is stored in the clean water tank, from where it is pumped into the distribution system for reuse (CH2M Hill Ltd 2006).

6.6.2.4 SMURRF Project Cost

Capital costs for the SMURRF were US$9 million. Of this, about US$6.3 million was for treatment and US$2.7 million for the distribution system. Approximately 12% of the plant cost, or US$750,000 was related to architectural components specifically designed to incorporate public art and education. One of the project's most expensive items was the concrete storage tank. The 1900 m^3 tanks for the raw water and the treated water were estimated at US$2 million. The tanks were expensive because of tight site conditions, the need to design one side for the tanks as a retaining wall for a freeway on-ramp and the architectural treatment (Antich *et al.* 2002).

6.6.3 Case Study in Germany (Potsdamer Platz)

6.6.3.1 Project Background

The DaimlerChrysler Area at the Potsdamer Platz, Berlin, was built under very strict stormwater management regulations to protect the local surface water. In October 1998, rainwater utilisation systems were introduced in Berlin as part of a large-scale urban re-development. Hence, the construction of the Postdamer Platz complex used an integrated ecological concept and included diverse sustainable techniques such as energy efficiency, environment-friendly building materials and a water-management system.

6.6.3.2 Design Requirement

In order to avoid the overloading of the existing mixed (or combined) sewerage system, the building permit issued by the city council stated that the new complex could drain rainwater into the sewerage system at a rate of no more than 3 L/ second/hectare, or 1% of flows during storm events. To comply with this regulation, the following techniques were implemented for the management of 23,000 m^3 rainwater which falls on 19 building complex per year: extensive and intensive green roofs;nd collection of roof-runoff for toilet flushing and plant irrigation (including intensive green roofs) and an artificial lake for rainwater retention and evaporation (Centgraf and Schmidt 2005).

6.6.3.3 Configuration of System

Under the re-development, a multilayered system of rainwater management and reuse has been created (see Figure 6.6). Rainwater is collected from green roofs of 19 buildings and the paved open areas. The water is piped for storage in two basement tanks with a total capacity of 3500 m^3. The storage capacity of the tanks corresponds to 15% of the annual precipitation (in Berlin 580 mm). This water is stored in the tanks except in peak events when it is drained over a weir to an artificial lake.

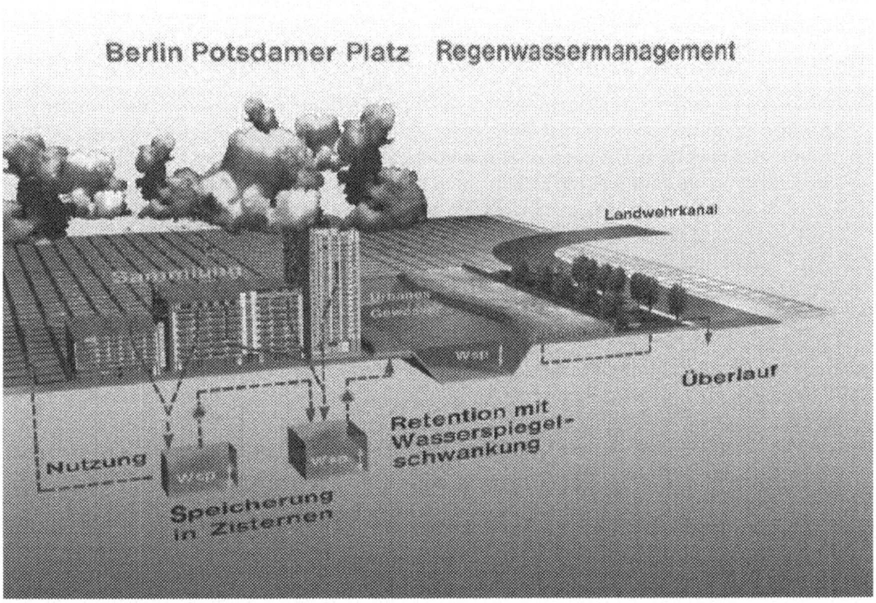

Figure 6.6 Postsdamer Platz multilayered rainwater management and reuse system. *Source*: Neckar 2003.

The artificial lake system covers a total area of 13,042 m^2 and has a volume of 15,000 m^3. It took 6100 m^3 of concrete, approximately 3-km piping and 3 km of cables to complete. The level of the artificial lake can be varied by 30 cm, which corresponds to an additional storage capacity of 17% of the average annual precipitation of the catchment area (Reichmann *et al.* 2004).

The water circulates continuously with maximum filtering capacity ranging from 30 to 150 m^3/hour for the different parts of the lake. Additionally, there are multilayered filters through which the water can be fed by three pumps in the control stations with a maximum combined capacity of 125 m^3/hour. The mechanism of the pumps and measuring devices are controlled by two programmable automatic systems (Siemens SIMATIC). The water eventually has a low nutrient concentration and can be pumped for toilet flushing, watering of green areas (including roofs with vegetative cover) and the replenishment of the artificial lake with a high transparency year round.

6.7 REFERENCES

Akan, A. O. and Houghtalen, R. J. 2003, Urban Hydrology, Hydraulics, and Stormwater Quality. John Wiley & Sons, Inc.

Antich, A., Gobas, H. R. and Salgaonkar, J. 2002, The Santa Monica Urban Runoff Recycling Facility and the Sustainable Environment. The Journal for Surface Water Quality Professionals 3[5] July/August.

Centgraf, C. S. and Schmidt, M. 2005, Water management to save energy, a decentralised approach to an integrated sustainable urban development. Proceedings of the International Conference on Rio5 World Climate and Energy Event, Brazil, February 15–17.

CH2M Hill Ltd. 2006, SMURRF Process Flow Description. http://santa-monica.org/epd/residents/Urban_Runoff/urban.htm.

Chui, P. C. 1991, Quality of stomwater runoff from an urbanised watershed. Environmental Monitoring and Assessment 19, 449–456.

Environmental Programs Division of Santa Monica. 2006, Clean Ocean Fact Sheet for Resident. http://santa-monica.org/EPD/printed_materials/pdf/UR_Residential_Final.pdf.

Hall, C. P. 2005, Municipal use of stormwater runoff. The Journal for Surface Water Quality Professionals 6[3] May/June.

Lee, M. F. and Haja, N. 1996, Collection of urban stormwater for potable water supply in Singapore. WQI May/June, 36–40.

Lim, C. H. and Lim, N. S. 1999, Urban stormwater collection for potable use. Water Supply 17[3/4], 503–509.

Mitchell, G., Mein, R. and McMahon, T. 1999, The Reuse Potential of Urban Stormwater and Wastewater. Cooperative Research Centre for Catchment Hydrology, Industry Report 99/14.

Neckar, M. L. 2003, Transportation, Urban Design, and the Environment: Highway 61/Red Rock Corridor. Research Report, Centre for Transport Study, University of Minnesota, MN.

Pitt, R., Field, R., Lalor, M. and Brown, M. 1995, Urban stormwater toxic pollutants: assessments, sources, and treatability. Water Environment Research 67[3], 260–275.

Reichmann, B., Teschner, K. and Schmidt, M. 2004, Tour a – innovative water concepts: process and rainwater management. Integrated runoff management in Berlin – also in areas without sewer systems. Proceedings of International Water Conference on Urban Waters: Problems and Solutions with Regard to Sustainability, Berlin, October 4–6.

Shapiro, N. 2003, The stranger amongst us: urban runoff, the forgotten local water resource. IProceedings of US EPA National Conference on Urban Stormwater: Enhancing Programs at the Local Level, Chicago, February 17–20.

United Nations Environment Programme. 2002, Rainwater harvesting and utilisation. An Environmentally Sound Approach for Sustainable Urban Water Management: An Introductory Guide for Decision-Makers. www.unep.or.jp/ietc/Publications/Urban/UrbanEnv-2/index.asp.

US Environmenatl Protection Agency (EPA). 2003, Protecting water quality from urban runoff. EPA 841-F-03-003.

7

Concluding Remarks

In Chapter 1, an introduction of the dual water supply systems has been given. The dual water supply systems involve two distribution systems, a fresh water system for potable use and another system conveying water for non-potable use, for example toilet flushing. First, the fresh water system is discussed as to its distribution and treatment. Then, the seawater distribution and its treatment are described. The use of seawater for toilet flushing in Hong Kong has undoubtedly helped reduce a considerable amount of fresh-water demand. Provided that fundamental precautions regarding the choice of materials are taken, a seawater supply system is technically no more difficult to construct, operate and maintain than any other reticulation systems.

There are problems and difficulties encountered in operating dual water supply systems, of course. Chapter 2 discusses these problems, namely chlorination problems, cross-connection problems, deposits and biological growth inside pipelines, corrosion problems, sudden deterioration of seawater quality, ecology problems and the treatment of mixed (salt + fresh) sewage. Each and every one of these difficulties is addressed, and it can be seen that based on the Hong Kong experience, there is no insurmountable problem in operating the dual water supply systems.

Chapter 3 makes a cost comparison of the dual water supply systems with the conventional single (potable) fresh water system. The net present value (NPV) is the tool used for comparison. In this chapter, first the concept of NPV and its application are introduced. Second, the capital cost, the running cost and the income generated from collecting fees from users are evaluated. The NPV method is then used to compare the existing dual water supply systems in Hong Kong and an imaginary single fresh-water supply system of a similar size. The result of the comparison shows that the single fresh-water supply system is nearly 40% more expensive than the existing dual water supply systems.

Normally, the dual water supply systems mean a fresh potable water supply system plus a seawater supply system. However, seawater can sometimes be substituted by reclaimed water. Chapter 4 describes the use of reclaimed water in the dual systems. Reclaimed water supply systems can be broadly classified by scale as municipal, district and building lot. On the municipal scale, the reclaimed water treatment plant is normally attached to a sewage treatment works, taking the secondarily treated effluent as the source water. On the district scale, if there is a sewage treatment works, the treated effluent can be taken as the source of reclaimed water, similar to that on the municipal scale. If primary sewage treatment only is provided, upgrading of the sewage treatment process is necessary for the production of reclaimed water. If there is no sewage treatment works in the district, the source water may be taken from the rainwater drainage system, which also receives wash water discharged from the buildings of the district, but not the foul water from toilets and the wash water from kitchens to avoid water treatment complications. On the scale of a single building lot, the sources of reclaimed water can be limited to the discharges from bathtubs or showers, hand-wash basins, washing machines and rainwater only. Recycled use of discharges from toilets and kitchens can also sometimes be considered. In such a case, the environmental impacts must be carefully assessed.

Seawater in the dual water supply systems can be substituted by reclaimed water as seen previously. It can, however, sometimes be substituted by raw fresh water (fresh water without rigorous treatment). Chapter 5 makes a cost comparison, using the NPV again, of seawater, raw fresh water and reclaimed water for toilet flushing. Four existing districts in Hong Kong, namely Southern District (SD) of Hong Kong Island, Sai Kung (SK), Northern New Territories (NNT) including Sheung Shui and Fanling and Northwest New Territories (NWNT) including Tin Shui Wai and Yuen Long, are not supplied with seawater for people living there to flush toilets. Therefore, these four districts are assumed to be supplied with either seawater, raw fresh water or reclaimed water for toilet flushing (imaginary cases again). First, three cases are considered:

Case 1 – toilet flushing system using seawater in SD, SK, NNT and NWNT

Case 2 – toilet flushing system using raw water in SD, SK, NNT and NWNT and

Case 3 – toilet flushing system using reclaimed water (treated effluent) in SD, SK, NNT and NWNT.

The cost comparison result shows that the dual water supply systems by adopting seawater for toilet flushing in these districts have the best engineering economy, then raw fresh water and then reclaimed water. Then, three more cases are also considered:

Case 4 – toilet flushing system using seawater in SD and SK, and raw water in NNT and NWNT

Case 5 – toilet flushing system using seawater in SD and SK, and reclaimed water (treated effluent) in NNT and NWNT and

Case 6 – toilet flushing system using seawater in SD and SK, raw water in NNT and reclaimed water (treated effluent) in NWNT.

The result shows that Case 4 is the cheapest, then Case 6, and then Case 5.

Chapter 6 discusses the possibility of using urban runoff as toilet flushing water. The chapter starts with describing the advantages of using urban runoff in the dual water supply systems. Then it goes on to describe the urban runoff collection system and then its water quality. It moves on to discuss the usual pollutants contained in urban runoff and the precautionary measures for protecting the quality of the runoff. Then, the chapter describes three cases of urban runoff collection, one in Singapore, one in the USA and one in Germany. Only the urban runoff collected in the USA and Germany cases are used for toilet flushing. Similar to reclaimed water, urban runoff collection systems can be broadly classified by scale as municipal/district and building lots. The Singapore and the USA cases are examples of municipal/district scale and the Germany case is of building lots' scale.

Chapter 7 contains concluding remarks. This is a short chapter summarising and discussing all main outcomes contained in the previous six chapters.

Index

IWA Publishing's authorised EU representative for General Product Safety Regulations is Diane D'Arras, 15 rue Duret, 75116 Paris, France, e-mail: safety@iwap.co.uk.

Printed and bound by CPI Group (UK) Ltd, Croydon, CR0 4YY
16/06/2026
02140357-0001

PERFORMING LAW

Actors, Affects, Spaces

Edited by

Peter Goodrich

Cardozo School of Law (Yeshiva University)

Anna Jayne Kimmel

Corcoran School of the Arts & Design, George Washington University

Bernadette Meyler

Stanford University

CAMBRIDGE
UNIVERSITY PRESS

CAMBRIDGE
UNIVERSITY PRESS

Shaftesbury Road, Cambridge CB2 8EA, United Kingdom

One Liberty Plaza, 20th Floor, New York, NY 10006, USA

477 Williamstown Road, Port Melbourne, VIC 3207, Australia

314–321, 3rd Floor, Plot 3, Splendor Forum, Jasola District Centre, New Delhi – 110025, India

Cambridge University Press is part of Cambridge University Press & Assessment, a department of the University of Cambridge.

We share the University's mission to contribute to society through the pursuit of education, learning and research at the highest international levels of excellence.

www.cambridge.org
Information on this title: www.cambridge.org/9781009674201

DOI: 10.1017/9781009674195

First published 2026

Cover image: Detail of a scene from the Commedia dell'arte by Arthur Keene (1930–2013) © Gary Watt.

A catalogue record for this publication is available from the British Library

Library of Congress Cataloging-in-Publication Data
Names: Goodrich, Peter, 1954– editor | Kimmel, Anna Jayne editor | Meyler, Bernadette editor
Title: Performing law : actors, affects, spaces / Peter Goodrich, Anna J Cardozo School of Law (Yeshiva University); Anna Jayne Kimmel, Corcoran School of the Arts & Design; Bernadette Meyler, Stanford University, California.
Description: Cambridge United Kingdom ; New York, NY : Cambridge University Press, 2026. | Includes bibliographical references and index.
Identifiers: LCCN 2025051592 | ISBN 9781009674201 hardback | ISBN 9781009674249 paperback | ISBN 9781009674195 ebook
Subjects: LCSH: Court proceedings | Procedure (Law) | Courts | Trials | Justice, Administration of | Judges
Classification: LCC K2110 .P47 2026
LC record available at https://lccn.loc.gov/2025051592

ISBN 978-1-009-67420-1 Hardback
ISBN 978-1-009-67424-9 Paperback

Contents

PART III TRANSGRESSIONS

Figures

Contributors

Başak Ertür teaches at Goldsmiths, University of London, in the Department of Visual Cultures. She is the author of *Spectacles and Spectres: A Performative Theory of Political Trials* (2023).

Peter Goodrich is a professor of law at Cardozo School of Law (Yeshiva University), and a visiting professor in the Division of Social Science, New York University Abu Dhabi. His most recent excursus is *Vision in Decision* (2023).

Piyel Haldar is a senior lecturer in law at the School of Law, Birkbeck College. He has published on the architecture of court buildings and on the wild energies of law as they flow through ley lines embedded in the natural landscape. He is currently writing a piece on the philosophical and theatrical absurdities of legal transmission as well as a monograph on the ecology of the common law.

Lorna Hutson teaches at Oxford and writes books on literature and law, including *The Invention of Suspicion* (2007), *The Oxford Handbook of English Law and Literature 1500–1700* (2017), and *England's Insular Imagining: The Elizabethan Erasure of Scotland* (2023), which won the Saltire Prize for best research book, 2025.

Anna Jayne Kimmel is an assistant professor of dance and affiliate faculty at the Nashman Center for Civic Engagement and Public Service at George Washington University. She is an associate editor at *Performance Research* and a board member of Performance Studies international. Her scholarship appears in *Dance Research Journal, Performance Research, Lateral, The Drama Review*, and *The Brooklyn Rail*, as well as various edited volumes.

Jisha Menon is the Robert G. Freeman Professor of International Studies at Stanford University, where she is a professor of theater and performance studies, and (by courtesy) of comparative literature and serves as the Fisher Family Director of Stanford Global Studies. She is the author of *Brutal Beauty: Aesthetics and Aspiration in Urban India* (2021) and *The Performance of Nationalism: India, Pakistan and the Memory of Partition* (2013).

Bernadette Meyler is the Carl and Sheila Spaeth Professor of Law at Stanford University. She is the author of *Theaters of Pardoning* and a coeditor of *The Oxford Handbook of Law and Humanities.*

Derek Miller is a professor of English at Harvard University, where he teaches courses in theater history and dramatic literature. His first monograph is *Copyright and the Value of Performance, 1770–1911* (2018). Visit visualizingbroadway.com for more information.

Subha Mukherji is a professor of early modern literature and culture at the University of Cambridge. She is the series editor of Crossroads of Knowledge in Early Modern Literature and most recently, with Dunstan Roberts, she edited *Literature and the Legal Imaginary: Knowing Justice.*

Julie Stone Peters is the H. Gordon Garbedian Professor of English and Comparative Literature at Columbia and affiliated faculty at Columbia Law School. Her most recent book is *Staging Witchcraft before the Law* (2024). A scholar of law and media history, she is currently writing about contemporary audiovisual legal cultures.

Simon Stern is a professor of law and English at the University of Toronto. He has published extensively on copyright, obscenity, search and seizure, legal fictions, and the history of legal methods. He is completing a study of Sherlock Holmes and the law.

Jesús R. Velasco is the Augustus R. Street Professor of Spanish and Portugese at Yale University. He has a PhD in philology from the University of Salamanca (1995) and a PhD in law from the École des hautes études en sciences sociales in Paris (2025). He has published, most recently, *Microliteratures: The Production of the Margin in Medieval and Early Modern Iberian Books* (2025), and *Dead Voice: Law, Philosophy, and Fiction* *in the Iberian Middle Ages* (2020). He is interested in the production of legal codes and legal thinking in the Mediterranean during the Middle Ages.

Marco Wan is a professor of law and the director of the Programme in Law and Literary Studies at the University of Hong Kong. He is the first legal scholar to be elected to the Hong Kong Academy of the Humanities. He serves as the managing editor of *Law & Literature.*